WITHDRAWN

The Calculus of Variations and Optimal Control
An Introduction

MATHEMATICAL CONCEPTS AND METHODS IN SCIENCE AND ENGINEERING

Series Editor: **Angelo Miele**
Mechanical Engineering and Mathematical Sciences, Rice University

Recent volumes in the series:

11 **INTEGRAL TRANSFORMS IN SCIENCE AND ENGINEERING** • *Kurt Bernardo Wolf*

12 **APPLIED MATHEMATICS:** An Intellectual Orientation • *Francis J. Murray*

14 **PRINCIPLES AND PROCEDURES OF NUMERICAL ANALYSIS**
   • *Ferenc Szidarovszky and Sidney Yakowitz*

16 **MATHEMATICAL PRINCIPLES IN MECHANICS AND ELECTROMAGNETISM,**
   Part A: Analytical and Continuum Mechanics • *C.-C. Wang*

17 **MATHEMATICAL PRINCIPLES IN MECHANICS AND ELECTROMAGNETISM,**
   Part B: Electromagnetism and Gravitation • *C.-C. Wang*

18 **SOLUTION METHODS FOR INTEGRAL EQUATIONS:** Theory and Applications
   • *Edited by Michael A. Golberg*

19 **DYNAMIC OPTIMIZATION AND MATHEMATICAL ECONOMICS** • *Edited by Pan-Tai Liu*

20 **DYNAMICAL SYSTEMS AND EVOLUTION EQUATIONS:** Theory and Applications
   • *J. A. Walker*

21 **ADVANCES IN GEOMETRIC PROGRAMMING** • *Edited by Mordecai Avriel*

22 **APPLICATIONS OF FUNCTIONAL ANALYSIS IN ENGINEERING** • *J. L. Nowinski*

23 **APPLIED PROBABILITY** • *Frank A. Haight*

24 **THE CALCULUS OF VARIATIONS AND OPTIMAL CONTROL:**
   An Introduction • *George Leitmann*

A Continuation Order Plan is available for this series. A continuation order will bring delivery of each new volume immediately upon publication. Volumes are billed only upon actual shipment. For further information please contact the publisher.

The Calculus of Variations and Optimal Control
An Introduction

George Leitmann
University of California
Berkeley, California

PLENUM PRESS · NEW YORK AND LONDON

Library of Congress Cataloging in Publication Data

Leitmann, George.
The calculus of variations and optimal control.

(Mathematical concepts and methods in science and engineering; v. 24)
Bibliography: p.
Includes index.
1. Calculus of variations. 2. Control theory. I. Title. II. Series.
QA315.L36   515'.64   81-4582
ISBN 0-306-40707-8   AACR2

First Printing — May 1981
Second Printing — September 1983
Third Printing — March 1986

© 1981 Plenum Press, New York
A Division of Plenum Publishing Corporation
233 Spring Street, New York, N.Y. 10013

All rights reserved

No part of this book may be reproduced, stored in a retrieval system, or transmitted, in any form or by any means, electronic, mechanical, photocopying, microfilming, recording, or otherwise, without written permission from the Publisher

Printed in the United States of America

QA   Leitmann, George.
315
.L36   The calculus of
1981   variations and optimal
       control
515.64 L536c

What is the answer?

        Alice B. Toklas

What is the question?

        Gertrude Stein

Preface

When the Tyrian princess Dido landed on the North African shore of the Mediterranean sea she was welcomed by a local chieftain. He offered her all the land that she could enclose between the shoreline and a rope of knotted cowhide. While the legend does not tell us, we may assume that Princess Dido arrived at the correct solution by stretching the rope into the shape of a circular arc and thereby maximized the area of the land upon which she was to found Carthage. This story of the founding of Carthage is apocryphal. Nonetheless it is probably the first account of a problem of the kind that inspired an entire mathematical discipline, the calculus of variations and its extensions such as the theory of optimal control.

This book is intended to present an introductory treatment of the calculus of variations in Part I and of optimal control theory in Part II. The discussion in Part I is restricted to the simplest problem of the calculus of variations. The topic is entirely classical; all of the basic theory had been developed before the turn of the century. Consequently the material comes from many sources; however, those most useful to me have been the books of Oskar Bolza and of George M. Ewing. Part II is devoted to the elementary aspects of the modern extension of the calculus of variations, the theory of optimal control of dynamical systems. Here the approach is not variational but rather geometric; it is based on a theory developed in collaboration with Austin Blaquière of the University of Paris.

This volume is the outgrowth of lecture notes for a course on the variational calculus and optimal control which has been taught at the University of California at Berkeley for over twenty years. Based on this experience, I believe that a first-year graduate student in an engineering or applied science curriculum should possess the requisite mathematical sophistication required for a reading of this text.

Over the years I have benefited greatly from fruitful discussions with many colleagues and students, too numerous to list here; they know who they are. However, two of them merit special mention. I am deeply grateful to Martin Corless and to Wolfram Stadler for their critical reading of the manuscript and for their constructive suggestions. I am also indebted to David G. Luenberger, William E. Schmitendorf, and Thomas L. Vincent for allowing me to quote from their work in Sections 13.12, 15.8, and 15.9 of the book.

George Leitmann

Contents

Symbols and Notation xi

PART I. CALCULUS OF VARIATIONS

1. Introduction 3
2. Problem Statement and Necessary Conditions for an Extremum........... 7
   - 2.1. Introduction 7
   - 2.2. The Simplest Problem with Fixed End Points 7
   - 2.3. Minima of Ordinary Functions 8
   - 2.4. Local Minima of Integrals 11
   - 2.5. The DuBois-Reymond Lemma 14
   - 2.6. The Necessary Condition of Euler 15
   - 2.7. Remarks 19
   - 2.8. Stationarity 22
   - *Exercises* 23
3. Integration of the Euler–Lagrange Equation 25
   - 3.1. Introduction 25
   - 3.2. The Function $f(\cdot)$ Independent of x: $f_x(t, x, r) \equiv 0$ 25
   - 3.3. The Function $f(\cdot)$ Independent of t: $f_t(t, x, r) \equiv 0$ 26
   - 3.4. The Function $f(\cdot)$ Independent of t and x: $f_t(t, x, r) \equiv 0$
     and $f_x(t, x, r) \equiv 0$ 27
   - 3.5. Examples 27
   - 3.6. Remarks 31
   - 3.7. The Function $f(\cdot)$ Linear in r: $f_{rr}(t, x, r) \equiv 0$ 33
   - *Exercises* 37
4. An Inverse Problem 39
   - 4.1. Problem Statement 39
   - 4.2. Solution 39

	4.3. Examples	41
	Exercises	44
5.	**The Weierstrass Necessary Condition**	**47**
	5.1. Introduction	47
	5.2. The Excess Function and a Necessary Condition	47
	5.3. Example	50
	5.4. The Legendre Necessary Condition	51
	5.5. Example	52
	Exercises	53
6.	**Jacobi's Necessary Condition**	**55**
	6.1. Introduction	55
	6.2. The Accessory Minimum Problem	55
	6.3. The Integration of Jacobi's Equation	60
	6.4. Example	64
	Exercises	65
7.	**Corner Conditions**	**67**
	7.1. Necessary Conditions	67
	7.2. Example	68
	Exercises	70
8.	**Concluding Remarks**	**71**

PART II. OPTIMAL CONTROL

9.	**Introduction**	77
10.	**Problem Statement and Optimality**	79
	10.1. Introduction	79
	10.2. Problem Statement	79
	10.3. Joining Controls and Additivity of Costs	83
	10.4. Optimal Cost and an Optimality Principle	85
	10.5. Augmented State Space and Trajectories	88
	10.6. Limiting and Optimal Isocost Surfaces	90
	10.7. Fundamental Properties of Trajectories and of Limiting Surfaces	92
	10.8. An Illustrative Example	95
	Exercises	98
11.	**Regular Optimal Trajectories**	**99**
	11.1. Regular Interior Points of a Limiting Surface	99
	11.2. Necessary Conditions at a Regular Interior Point	105
	11.3. A Linear Transformation	107

Contents   xi

    11.4. Transformation of the Tangent Plane 109
    11.5. Regular Optimal Trajectories 111
    11.6. An Illustrative Example 114
    11.7. The Terminal Transversality Condition 115
    11.8. A Maximum Principle 118
    11.9. Remarks 119
    11.10. Extremal Control 121
    11.11. An Illustrative Example 122
    Exercises 123

12. **Examples of Extremal Control** 125
    12.1. Time-Optimality for a Constant-Power Rocket 125
    12.2. A Problem of Time-Optimal Navigation 130
    12.3. The Minimum Distance to a Given Curve 135
    Exercises 138

13. **Some Generalizations** 139
    13.1. Introduction 139
    13.2. The Initial Transversality Condition 140
    13.3. Minimum Distance between Given Curves 144
    13.4. A Terminal Cost 145
    13.5. An Extremal Thrust Control for a Rocket 150
    13.6. Nonautonomous Systems 155
    13.7. Problems with a Fixed Interval 160
    13.8. A Minimum Fuel Rendezvous of a Constant-Power Rocket 161
    13.9. The Simplest Problem of the Calculus of Variations 164
    13.10. State-Dependent Control Constraints 167
    13.11. A Time-Optimal Regulator with Velocity-Dependent Control
           Bounds 174
    13.12. Isoperimetric Constraints 181
    13.13. Dido's Problem 188
    13.14. Parameter Optimization 190
    13.15. An Illustrative Example 195
    13.16. End Point Inequality Constraints 196
    13.17. An Illustrative Example 198
    Exercises 205

14. **Special Systems** 211
    14.1. Introduction 211
    14.2. Linear Time-Invariant State Equations 211
    14.3. The Switching Function 212
    14.4. Time-Optimality and Bang–Bang Control 213
    14.5. The Number of Switches 215
    14.6. A One-Dimensional Time-Optimal Regulator 217
    14.7. A Three-Dimensional Time-Optimal Regulator 222
    14.8. Singular Control 225
    14.9. The Maximum Range of a Thrust-Limited Rocket 226

	14.10. The Maximum Range of a Rocket in Horizontal Flight	231
	Exercises	237
15.	**Sufficient Conditions**	**241**
	15.1. Introduction	241
	15.2. A Field Theorem	241
	15.3. An Illustrative Example	244
	15.4. Another Field Theorem	245
	15.5. Time-Optimality for a Constant-Power Rocket	247
	15.6. A Direct Sufficiency Theorem	249
	15.7. An Illustrative Example	251
	15.8. Life History Strategies of Plants	252
	15.9. An Economic Control Problem	256
	Exercises	262
16.	**Feedback Control**	**265**
	16.1. Introduction	265
	16.2. The Synthesis of Optimal Feedback Control	268
	16.3. The Linear-Quadratic Problem	269
	16.4. The Existence of Feedback Solutions	275
	16.5. An Illustrative Example	278
	Exercises	283
17.	**Optimization with Vector-Valued Cost**	**285**
	17.1. Introduction	285
	17.2. How to Choose a Cheese	286
	17.3. Pareto-Optimal Control	292
	17.4. Necessary Conditions for Pareto-Optimality	293
	17.5. Sufficient Conditions for Pareto-Optimality	295
	17.6. An Illustrative Example	296
	Exercises	298

REFERENCES ... 301
BIBLIOGRAPHY ... 305
INDEX ... 309

Symbols and Notation

Standard mathematical symbols and notation are used in this book. The most commonly used symbols are defined first. Thereafter we give the definitions of the basic notation employed in the text.

Symbols

$\triangleq$	equals by definition; denotes
$=$	equals, is equivalent to
$\neq$	does not equal; is not equivalent to
$\equiv$	equals identically; is the same as
$\not\equiv$	does not equal identically; is not the same as
$\leq$ ($\geq$)	is less (greater) than or equal to
$<$ ($>$)	is less (greater) than
$\forall$	for all, for every
$\in$	is an element (member) of; belongs to
$\notin$	is not an element (member) of; does not belong to
$\emptyset$	empty set
$\subset$	is a subset of; is contained in
$\supset$	contains
$\cup$	union
$\cap$	intersection
$\times$	Cartesian product
$\{e \mid P\}$	set of all e having property P
R^1	set of real numbers; real line
$[a, b]$	$\{x \in R^1 \mid a \leq x \leq b\}$

(a, b)	$\{x \in R^1 \mid a < x < b\}$
$(a, b]$	$\{x \in R^1 \mid a < x \le b\}$
$[a, b)$	$\{x \in R^1 \mid a \le x < b\}$
$\setminus$	subtraction of sets; that is, $A \setminus B \triangleq \{e \mid e \in A, e \notin B\}$
$\lvert\ \rvert$	absolute value
$\lVert\ \rVert$	Euclidean norm
inf (sup)	infimum (supremum)
min (max)	minimum (maximum)
sgn x	signum; that is, for $x \in R^1$, sgn $x = 1$ if $x > 0$, sgn $x = -1$ if $x < 0$
T	transpose (superscript)

Spaces

The set of all ordered n-tuples of real numbers is denoted by R^n; that is,

$$R^n \triangleq R^1 \times R^1 \times \cdots \times R^1 \quad (n \text{ times}).$$

Thus, given an ordered n-tuple of real numbers, $x_1, x_2, \ldots, x_n$, we consider it to be a *vector* $x \in R^n$. We let all vectors be column vectors; that is,

$$x \triangleq \begin{bmatrix} x_1 \\ x_2 \\ \vdots \\ x_n \end{bmatrix} = [x_1\ x_2\ \cdots\ x_n]^T.$$

By endowing R^n with the *natural basis* $\{e^1, e^2, \ldots, e^n\}$, where $e^i \in R^n$ and

$$e^{iT} e^j = \delta_{ij}, \qquad \delta_{ij} \triangleq \begin{cases} 1 & \text{if } i = j \\ 0 & \text{if } i \ne j, \end{cases}$$

we assure that R^n is a Euclidean space. In particular it follows that

$$x^T x = \sum_{i=1}^n x_i^2 = \lVert x \rVert^2.$$

The *closure of a set* $X \subset R^n$, denoted by $\overline{X}$, is the set together with all of its accumulation points; that is,

$$\overline{X} \triangleq X \cup \{x \mid \text{there is a sequence } x_i \in X, i = 1, 2, \ldots,$$
$$\text{such that } x_i \text{ converges to } x\}.$$

Symbols and Notation

Functions

Given the nonempty sets X and Y,

$$f(\cdot): X \to Y$$

denotes a *function* (mapping) from the domain X into the range Y; that is, associated with $x \in X$ there is one and only one $y \in Y$. We write $y = f(x)$, and we term $f(x)$ the value of the function at x.

The scalar-valued function $f(\cdot):[a,b] \to R^1$, $[a,b] \subset R^1$, $a<b$, is of *class* C^k if and only if it and its first k derivatives are continuous on $[a,b]$. Such a function of class C^1 is also called *smooth*.

The scalar-valued function $f(\cdot):[a,b] \to R^1$, $[a,b] \subset R^1$, $a<b$, is *piecewise continuous* if and only if it is continuous on $[a,b]$ with the exception of a finite number of points of (a,b) where it possesses defined left and right limits; that is, if $f(\cdot)$ is discontinuous at $\bar{x} \in (a,b)$, then

$$f(\bar{x}-0) \triangleq \lim_{\substack{x \to \bar{x} \\ x < \bar{x}}} f(x)$$

and

$$f(\bar{x}+0) \triangleq \lim_{\substack{x \to \bar{x} \\ x > \bar{x}}} f(x)$$

are defined. In order to have $f(x)$ defined for all $x \in [a,b]$, we take

$$f(a) = f(a+0),$$
$$f(b) = f(b-0),$$

and if $f(\cdot)$ is discontinuous at $\bar{x} \in (a,b)$ we take

$$f(\bar{x}) = f(\bar{x}-0)$$

or

$$f(\bar{x}) = f(\bar{x}+0).$$

The function $f(\cdot):[a,b] \to R^1$, $[a,b] \subset R^1$, $a<b$, is *piecewise smooth* if and only if it is continuous and its first derivative is piecewise continuous on $[a,b]$. If the first derivative is discontinuous at $\bar{x} \in (a,b)$, then the point $(x,y) = (\bar{x}, f(\bar{x}))$ is termed a *corner* of $f(\cdot)$.

The same notation is used for a vector-valued function $f(\cdot):[a,b]\to R^n$, $[a,b]\subset R^1$, $a<b$, provided the appropriate conditions are satisfied by its components which are scalar-valued; for instance, $f(\cdot):[a,b]\to R^n$ is of class C^k if and only if the functions $f_i(\cdot):[a,b]\to R^1$, $i=1,2,\ldots,n$, are of class C^k, where $f(x) \triangleq [f_1(x)\ f_2(x)\cdots f_n(x)]^T$.

The function $f(\cdot): X\to R^1$, $X\subset R^m$, is of class C^k if and only if it and its partial derivatives up to and including order k are continuous on X.

Given a function $f(\cdot): X\to Y$ and $Z\subset X$, the *restriction* of $f(\cdot)$ to Z, denoted by $f(\cdot)|_Z$, is the function $f(\cdot)|_Z: Z\to Y$ such that $f(x)|_Z = f(x)$ for all $x\in Z$.

Consider the function $o(\cdot):[a,b]\to R^n$, $[a,b]\subset R^1$, $a<0$, $b>0$, such that

$$\lim_{x\to 0}\frac{o(x)}{x}=0.$$

Every function having this property is denoted by $o(\cdot)$.

The function $f(\cdot):[a,b]\to R^1$, $[a,b]\subset R^1$, is *convex* if and only if, for every x and y in $[a,b]$ and for every $\alpha\in[0,1]$, we have

$$f(z)\leq \alpha f(y)+(1-\alpha)f(x),$$

where

$$z=\alpha y+(1-\alpha)x.$$

Finally, given a function $f(\cdot): X\to R^1$, $X\subset R^n$, that is differentiable at $x\in X$, we let

$$\operatorname{grad} f(x) \triangleq \left[\frac{\partial f(x)}{\partial x_1}\ \frac{\partial f(x)}{\partial x_2}\ \cdots\ \frac{\partial f(x)}{\partial x_n}\right]^T.$$

Part I

Calculus of Variations

1

Introduction

All of us "know" the answer to the question: What is the shape of the shortest plane curve connecting two given points? Of course it is a straight line. In mathematical terms one may pose this question as follows. Consider the family of all piecewise smooth functions

$$y(\cdot):[x_0, x_1] \to R^1, \qquad x_0 < x_1,$$

satisfying

$$y(x_0) = y_0, \qquad y(x_1) = y_1,$$

where x_0, x_1, y_0, and y_1 are prescribed. Find a function $y^*(\cdot)$ in the family defined above that yields the curve of minimum length joining points (x_0, y_0) and (x_1, y_1). For given $y(\cdot)$, the length of the curve is

$$\int_{x_0}^{x_1} \left[1 + (dy/dx)^2\right]^{1/2} dx \qquad (1.1)$$

so that we seek $y^*(\cdot)$ such that

$$\int_{x_0}^{x_1} \left[1 + (dy^*/dx)^2\right]^{1/2} dx \leq \int_{x_0}^{x_1} \left[1 + (dy/dx)^2\right]^{1/2} dx \qquad (1.2)$$

for all $y(\cdot)$ in the class specified above.

A less trivial and considerably more difficult problem is that of determining the thrust program which results in maximizing the flight distance or range of a rocket plane in horizontal flight; see also Exercise 3.6

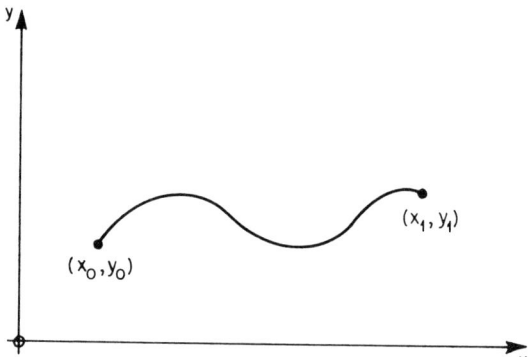

Figure 1.1. A curve connecting given points.

and Section 14.10. Let

$t=$ time,
$x=$ horizontal distance (range),
$v=$ speed,
$m=$ mass,
$T=-c\,dm/dt=$ thrust, $c=$ constant >0,
$L=$ lift,
$D=$ drag.

We assume that the lift, L, is adjusted to balance the weight, mg, $g=$ constant >0, so that the rocket moves horizontally. We assume further that the aerodynamic drag depends on the speed and lift as follows (Ref. 1.1):

$$D=Av^2+BL^2, \quad A \text{ and } B=\text{constants}>0.$$

Then the equations of motion of the rocket are

$$\frac{dx}{dt}=v,$$
$$m\frac{dv}{dt}=c\beta-D, \quad D=Av^2+Bg^2m^2, \quad (1.3)$$
$$\frac{dm}{dt}=-\beta.$$

Now, given the initial and terminal values of the speed v and mass m (and

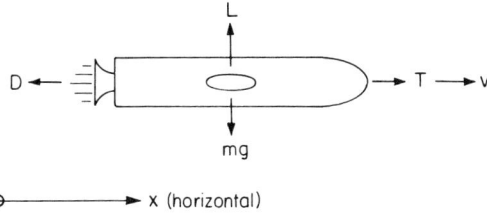

Figure 1.2. Forces acting on a rocket.

hence of the fuel consumed), we wish to determine the thrust program—that is, how T must be varied—in order to maximize the range $x(t_1)-x(0)$.

From (1.3) one obtains

$$dx = -\frac{cv}{D}\left[dm + \frac{m}{c}dv\right]$$

so that the range is

$$x(t_1) - x(0) = \int_{m_1}^{m_0} \frac{cv}{D}\left[1 + \frac{m}{c}\frac{dv}{dm}\right]dm \qquad (1.4)$$

where $m(0)=m_0$, $m(t_1)=m_1<m_0$, $v(0)=v_0$ and $v(t_1)=v_1$ are prescribed. We may now restate the problem more simply; namely, determine the speed v as a function of the mass m, satisfying the given end conditions and maximizing the value of the integral (1.4). From (1.3) one then has

$$T = c\beta = \frac{cD}{c + m\, dv/dm} \qquad (1.5)$$

yielding the thrust program as a function of m and v, and thence of m.

These examples typify the simplest problems of the calculus of variations. In the next chapter we state such problems in a general way.

2

Problem Statement and Necessary Conditions for an Extremum

2.1. Introduction

This chapter is devoted to a discussion of the so-called "simplest problem with fixed endpoints." Loosely speaking it is the problem of extremizing—that is, minimizing or maximizing—the value of an integral which depends on the choice of a function in a given class of functions, in particular, functions whose values at the endpoints of a given interval are prescribed. We shall begin by making these notions more precise. The remainder of the chapter is concerned with a condition that must be satisfied by a function that extremizes the value of an integral.

2.2. The Simplest Problem with Fixed End Points

Let $[t_0, t_1]$, $t_0 < t_1$, be a given bounded interval of the real line R^1. Let $\mathscr{X}$ denote the class of all piecewise smooth functions $x(\cdot):[t_0, t_1] \to R^1$ which satisfy the end conditions

$$x(t_0) = x_0, \quad x(t_1) = x_1, \tag{2.1}$$

where x_0 and x_1 are prescribed. We call $\mathscr{X}$ the set of *admissible* functions.

Consider also a given function $f(\cdot):[t_0, t_1] \times R^2 \to R^1$ whose values we denote by $f(t, x, r)$ for $t \in [t_0, t_1]$, $x \in R^1$, $r \in R^1$. Henceforth we shall assume that $f(\cdot)$ possesses as many partial derivatives as may be required at any stage of our discussion. We shall denote these partial derivatives by appropriate subscripts; for example,

$$f_t \triangleq \frac{\partial f}{\partial t}, \quad f_x \triangleq \frac{\partial f}{\partial x}, \quad f_r \triangleq \frac{\partial f}{\partial r}, \quad f_{tt} \triangleq \frac{\partial^2 f}{\partial t^2}, \quad f_{tx} \triangleq \frac{\partial^2 f}{\partial t \partial x}, \quad \text{etc.}$$

Now consider the function $J(\cdot): \mathcal{X} \to R^1$ whose values are

$$J(x(\cdot)) \triangleq \int_{t_0}^{t_1} f[t, x(t), \dot{x}(t)] \, dt, \quad x(\cdot) \in \mathcal{X} \tag{2.2}$$

where $\dot{x} \triangleq dx/dt$. Such a function of a function is sometimes called a functional. The problem to be discussed is that of finding a function—or functions—in the class of admissible functions $\mathcal{X}$, which extremizes the value of integral $J(\cdot)$.

The function $x^*(\cdot) \in \mathcal{X}$ renders the *global (absolute) minimum* of $J(\cdot)$ with respect to all $x(\cdot) \in \mathcal{X}$ if and only if

$$J(x^*(\cdot)) \le J(x(\cdot)) \quad \forall x(\cdot) \in \mathcal{X}.$$

The *global maximum* is defined in the analogous fashion by reversing the inequality. Henceforth we shall restrict the discussion to minima. This does not lead to any loss of generality since

$$-J(x^*(\cdot)) \ge -J(x(\cdot)) \quad \forall x(\cdot) \in \mathcal{X};$$

that is, a function $x^*(\cdot)$ which minimizes $J(\cdot)$ also maximizes $-J(\cdot)$. Other types of minima will be introduced subsequently.

Before continuing the discussion of global minima of integrals let us recall some facts about ordinary functions, that is, functions of a variable; such functions are sometimes called point functions to distinguish them from functionals. The results about minima of ordinary functions will be given without proof; the relevant proofs can be found in any standard text on the calculus (for instance, Refs. 2.1–2.3).

2.3. Minima of Ordinary Functions

Consider a function $\phi(\cdot): [\xi_0, \xi_1] \to R^1$. If $\phi(\cdot)$ is continuous on the bounded (and closed) interval $[\xi_0, \xi_1]$, then there is a $\xi^* \in [\xi_0, \xi_1]$ such that

$$\phi(\xi^*) \le \phi(\xi) \quad \forall \xi \in [\xi_0, \xi_1]. \tag{2.3}$$

Chap. 2 • Problem Statement

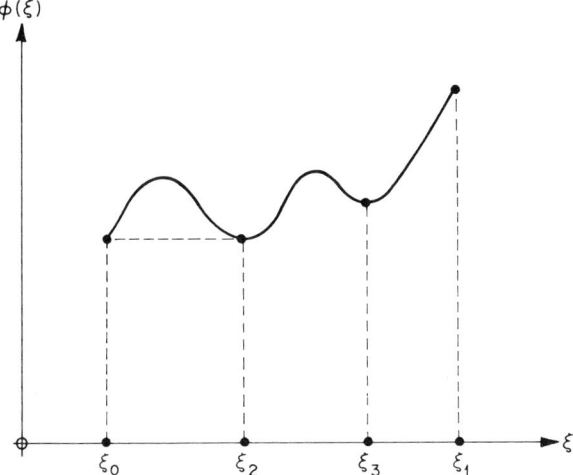

Figure 2.1. Minima of a function.

There may be other values of ξ for which $\phi(\cdot)$ attains a global minimum on $[\xi_0, \xi_1]$; that is, ξ^* need not be unique. However, the global minimum of $\phi(\cdot)$, namely, $\phi(\xi^*)$, is unique. The function $\phi(\cdot)$ in Figure 2.1 possesses a global minimum at two values of ξ—at $\xi = \xi_0$ and at $\xi = \xi_2$.

Let $N(\delta, \xi^*)$ denote a δ-*neighborhood* of ξ^*; that is,

$$N(\delta, \xi^*) \triangleq \{\xi \in R^1 | |\xi - \xi^*| < \delta\}.$$

If there is a $\delta > 0$ such that

$$\phi(\xi^*) \leq \phi(\xi) \quad \forall \xi \in [\xi_0, \xi_1] \cap N(\delta, \xi^*), \tag{2.4}$$

then ξ^* furnishes a *local (relative) minimum* of $\phi(\cdot)$. It is an immediate consequence of the definitions of global and local minima that a global minimum is always a local one; the converse need not be true. For example, in Figure 2.1, ξ_0, ξ_2, and ξ_3 furnish local minima, but ξ_3 does not render a global minimum.

If $\phi(\cdot)$ is twice differentiable on the open interval (ξ_0, ξ_1) and if $\phi(\xi^*)$ is a local minimum at $\xi^* \in (\xi_0, \xi_1)$, then

$$\phi'(\xi^*) = 0, \quad \phi''(\xi^*) \geq 0, \tag{2.5}$$

where $\phi' \triangleq d\phi/d\xi$, $\phi'' \triangleq d^2\phi/d\xi^2$. The condition (2.5) is *necessary*, but not

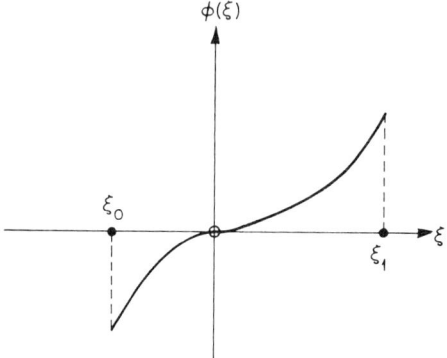

Figure 2.2. $\phi(\xi)=\xi^3$.

sufficient, for a local minimum. For example, consider $\phi(\xi)=\xi^3$ with $\xi\in[\xi_0,\xi_1]$ and $\xi_0<0$, $\xi_1>0$; see Figure 2.2. Thus, even though the condition (2.5) is satisfied at $\xi=0$, $\phi(0)$ is not a local minimum.

However, if

$$\phi'(\xi^*)=0, \qquad \phi''(\xi^*)>0 \tag{2.6}$$

for $\xi^*\in(\xi_0,\xi_1)$, then $\phi(\xi^*)$ is a local minimum; that is, this condition is a *sufficient* condition for a *local* minimum.

If $\phi(\cdot)$ is *convex*, and in particular if

$$\phi'(\xi^*)=0 \text{ and } \phi''(\xi)\geq 0 \qquad \forall \xi\in[\xi_0,\xi_1], \tag{2.7}$$

then $\phi(\xi^*)$ is a *global* minimum. This is illustrated in Figures 2.3a and 2.3b.

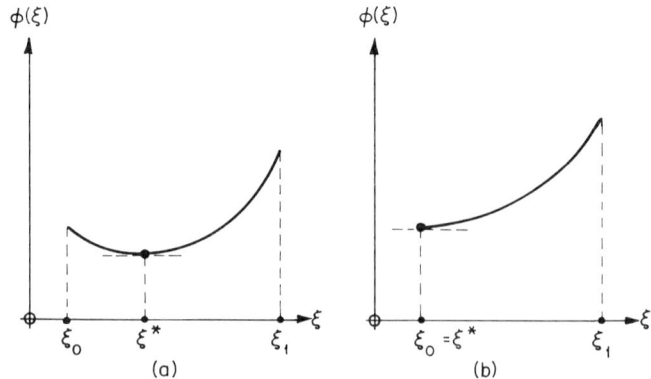

Figure 2.3. Minima of convex functions.

Having defined the notions of global as well as of local minima for ordinary functions, let us see how the latter can be defined for functionals given by (2.2).

2.4. Local Minima of Integrals

In Section 2.2 we defined the global minimum of $J(\cdot)$. To define *local* minima we must introduce the idea of a "neighborhood" in the space of piecewise smooth functions. This, in turn, requires the notion of "distance" between functions. We introduce two kinds of distance.

Consider two functions, $x(\cdot):[t_0, t_1] \to R^1$ and $\bar{x}(\cdot):[t_0, t_1] \to R^1$, which are piecewise smooth. Thus their derivatives may be discontinuous at a finite number of points of (t_0, t_1). Let D denote the set of points of (t_0, t_1) where either $\dot{x}(\cdot)$ or $\dot{\bar{x}}(\cdot)$ has a discontinuity.

The *distance of order* 0 between $x(\cdot)$ and $\bar{x}(\cdot)$ is

$$d_0[x(\cdot), \bar{x}(\cdot)] \stackrel{\triangle}{=} \sup_{t \in [t_0, t_1]} |x(t) - \bar{x}(t)|. \tag{2.8}$$

The *distance of order* 1 between $x(\cdot)$ and $\bar{x}(\cdot)$ is

$$d_1[x(\cdot), \bar{x}(\cdot)] \stackrel{\triangle}{=} d_0[x(\cdot), \bar{x}(\cdot)] + \sup_{t \in [t_0, t_1] \setminus D} |\dot{x}(t) - \dot{\bar{x}}(t)|. \tag{2.9}$$

Thus the functions $x(\cdot)$ and $\bar{x}(\cdot)$ are "near" each other in the sense of zero-order distance if their values are close to each other at every $t \in [t_0, t_1]$. They are "near" each other in the sense of first-order distance if, in addition, their slopes are close to each other at every $t \in [t_0, t_1]$ where they are defined.

Now we are ready to define neighborhoods. Given a piecewise smooth function $x(\cdot):[t_0, t_1] \to R^1$ and a positive $\delta \in R^1$, a δ-*neighborhood of order* 0 of $x(\cdot)$ is

$$N_0[\delta, x(\cdot)] \stackrel{\triangle}{=} \{\text{piecewise smooth } \bar{x}(\cdot):[t_0, t_1] \to R^1 | d_0[x(\cdot), \bar{x}(\cdot)] < \delta\}, \tag{2.10}$$

while a δ-*neighborhood of order* 1 of $x(\cdot)$ is

$$N_1[\delta, x(\cdot)] \stackrel{\triangle}{=} \{\text{piecewise smooth } \bar{x}(\cdot):[t_0, t_1] \to R^1 | d_1[x(\cdot), \bar{x}(\cdot)] < \delta\}. \tag{2.11}$$

Note that $\bar{x}(\cdot) \in N_1[\delta, x(\cdot)]$ implies that $\bar{x}(\cdot) \in N_0[\delta, x(\cdot)]$, but not conversely.

Now we are able to define two types of *local* minima of $J(\cdot)$. The function $x^*(\cdot) \in \mathcal{X}$ furnishes a *strong local minimum* of $J(\cdot)$ on $\mathcal{X}$ if and only if there is a $\delta > 0$ such that

$$J(x^*(\cdot)) \leq J(x(\cdot)) \qquad \forall x(\cdot) \in \mathcal{X} \cap N_0[\delta, x^*(\cdot)]. \tag{2.12}$$

The function $x^*(\cdot) \in \mathcal{X}$ furnishes a *weak local minimum* of $J(\cdot)$ on $\mathcal{X}$ if and only if there is a $\delta > 0$ such that

$$J(x^*(\cdot)) \leq J(x(\cdot)) \qquad \forall x(\cdot) \in \mathcal{X} \cap N_1[\delta, x^*(\cdot)]. \tag{2.13}$$

In the definitions of global, strong local, and weak local minima, the function $x^*(\cdot)$ is compared to members of successively smaller sets of functions; see Figure 2.4. Thus we conclude that a global minimum $\Rightarrow$ a strong local minimum $\Rightarrow$ a weak local minimum. Consequently a condition that is necessary for a weak local minimum is necessary for a strong local minimum. In turn, a condition that is necessary for a strong local minimum is necessary for a global minimum.

It is instructive to give examples of functions which belong to δ-neighborhoods of order 1 and 0, respectively, of a given function $x^*(\cdot)$. Let $\eta(\cdot): [t_0, t_1] \to R^1$ be piecewise smooth, and let $\varepsilon = $ constant, $\varepsilon \in [-\bar{\varepsilon}, \bar{\varepsilon}]$, where $\bar{\varepsilon} = $ constant > 0 such that

$$\sup_{t \in [t_0, t_1]} \bar{\varepsilon} |\eta(t)| + \sup_{t \in [t_0, t_1] \setminus D} \bar{\varepsilon} |\dot{\eta}(t)| < \delta$$

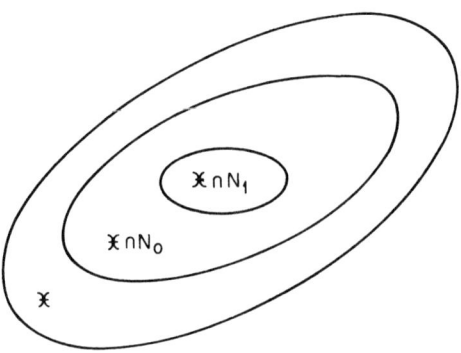

Figure 2.4. Sets of functions.

for a given $\delta = $ constant > 0. Then a function $x(\cdot): [t_0, t_1] \to R^1$ with values

$$x(t) = x^*(t) + \varepsilon \eta(t), \qquad t \in [t_0, t_1],$$

belongs to the δ-neighborhood of order 1 of $x^*(\cdot)$; that is,

$$x(\cdot) \in N_1[\delta, x^*(\cdot)].$$

It is a little more difficult to construct a function that belongs to a δ-neighborhood of order 0 but not to one of order 1. Given the function $x^*(\cdot): [t_0, t_1] \to R^1$, consider the function $x(\cdot): [t_0, t_1] \to R^1$ whose values are

$$x(t) = \begin{cases} x^*(t) & \text{for } t \in [t_0, a] \cup [b, t_1] \\ X(t) & \text{for } t \in [a, \varepsilon] \\ \phi(t, \varepsilon) & \text{for } t \in [\varepsilon, b] \end{cases}$$

where

$$X(t) \triangleq x^*(a) + q(t - a),$$

$$\phi(t, \varepsilon) \triangleq x^*(t) + [X(\varepsilon) - x^*(\varepsilon)] \frac{b - t}{b - \varepsilon}$$

and a, b, ε, and q are constants such that a is not the abscissa of a point of discontinuity of $\dot{x}^*(t)$, $t_0 \leq a < \varepsilon < b \leq t_1$, and $q \in R^1$. An illustration of such

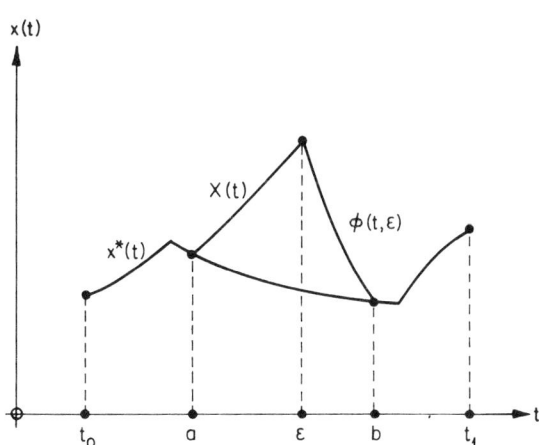

Figure 2.5. A function $x(\cdot) \in N_0[\delta, x^*(\cdot)]$.

a function is shown in Figure 2.5. Thus it follows that

$$x(t) \to x^*(t) \quad \text{as } \varepsilon \to a,$$

but in general it is not true that

$$\dot{x}(a+0) = \dot{X}(a) \to \dot{x}^*(a) \quad \text{as } \varepsilon \to a.$$

Clearly, given $\delta > 0$, for any a, b, and q, one can choose ε such that

$$d_0[x^*(\cdot), x(\cdot)] < \delta.$$

We shall return to these functions subsequently in deriving necessary conditions for an extremum of $J(\cdot)$. Now we turn to another preliminary result.

2.5. The DuBois-Reymond Lemma

In preparation for deriving a necessary condition for a weak local minimum of $J(\cdot)$, we require the following lemma (named after its discoverer).[†]

Lemma 2.1. *If $m(\cdot): [t_0, t_1] \to R^1$ is a given piecewise continuous function and if*

$$\int_{t_0}^{t_1} m(t) \dot{\eta}(t) \, dt = 0 \tag{2.14}$$

for all piecewise smooth functions $\eta(\cdot): [t_0, t_1] \to R^1$ such that $\eta(t_0) = \eta(t_1) = 0$, then $m(t) = \text{constant} \triangleq c$ on $[t_0, t_1]$.

Proof. Since it is desired that $m(t) = c$ for all $t \in [t_0, t_1]$, let us define a constant

$$c \triangleq \frac{1}{t_1 - t_0} \int_{t_0}^{t_1} m(t) \, dt. \tag{2.15}$$

[†] DuBois-Reymond published this lemma in 1879.

Also, in view of the hypothesis that $\eta(t_0)=\eta(t_1)$,

$$\int_{t_0}^{t_1} c\dot{\eta}(t)\,dt = c[\eta(t_1)-\eta(t_0)] = 0$$

so that, by hypothesis (2.14),

$$\int_{t_0}^{t_1} [m(t)-c]\dot{\eta}(t)\,dt = 0. \tag{2.16}$$

Since (2.16) must hold for all piecewise smooth $\eta(\cdot)$, consider an $\eta(\cdot)$ with values

$$\eta(t) = \int_{t_0}^{t} [m(\tau)-c]\,d\tau. \tag{2.17}$$

This function is piecewise smooth since its derivative, $\dot{\eta}(t) = m(t)-c$, is piecewise continuous. Clearly $\eta(t_0)=0$ and, as a consequence of (2.15) and (2.17), $\eta(t_1)=0$. Moreover, on substituting for $\dot{\eta}(t)$ in (2.16), one has

$$\int_{t_0}^{t_1} [m(t)-c]^2\,dt = 0. \tag{2.18}$$

Let $\hat{t} \in [t_0, t_1]$ be a point where $m(\cdot)$ is continuous. If $m(\hat{t}) \neq c$, then there is a subinterval of positive length, containing $\hat{t}$, on which $m(t) \neq c$; but then (2.18) cannot hold. Thus, $m(t)=c$ at all points of continuity, and hence also at possible discontinuity points (where it equals the limit from below or from above); hence, $m(t)=c$ for all $t \in [t_0, t_1]$. □

Now we are ready to derive a condition that must be satisfied by a function that results in the minimum value of an integral.

2.6. The Necessary Condition of Euler

As mentioned earlier, a *global* minimum is also a *weak local* minimum. Thus a *necessary* condition for a weak local minimum is also necessary for a global minimum. The following theorem, named after Leonhard Euler,[†] is concerned with such a necessary condition.

[†]Swiss mathematician, 1707–1783.

Theorem 2.1. *If $x^*(\cdot)$ furnishes a weak local minimum of $J(\cdot)$ on $\mathcal{X}$, then there is a constant c such that*

$$f_{\dot{x}}[t, x^*(t), \dot{x}^*(t)] = \int_{t_0}^{t} f_x[\tau, x^*(\tau), \dot{x}^*(\tau)] \, d\tau + c \qquad (2.19)$$

for all $t \in [t_0, t_1]$. At points of discontinuity t of $\dot{x}^*(\cdot)$ this condition applies for $\dot{x}^*(t-0)$ and $\dot{x}^*(t+0)$, respectively.

Proof. Let $\eta(\cdot):[t_0, t_1] \to R^1$ be any piecewise smooth function that satisfies the end conditions $\eta(t_0) = \eta(t_1) = 0$. Thus, given an $\varepsilon \in R^1$, the function $x^*(\cdot) + \varepsilon \eta(\cdot) \in \mathcal{X}$. Furthermore, $x^*(t) + \varepsilon \eta(t) \to x^*(t)$ and $\dot{x}^*(t) + \varepsilon \dot{\eta}(t) \to \dot{x}^*(t)$ as $\varepsilon \to 0$. Hence, given a $\delta > 0$, $x^*(\cdot) + \varepsilon \eta(\cdot) \in \mathcal{X} \cap N_1[\delta, x^*(\cdot)]$ for sufficiently small $|\varepsilon|$. Consequently, if $x^*(\cdot)$ renders at least a weak local minimum of $J(\cdot)$, then

$$J(x^*(\cdot)) \le J(x^*(\cdot) + \varepsilon \eta(\cdot)) \qquad \forall x^*(\cdot) + \varepsilon \eta(\cdot) \in \mathcal{X} \cap N_1[\delta, x^*(\cdot)]. \qquad (2.20)$$

Consider the function $F(\cdot): R^1 \to R^1$ with values

$$F(\varepsilon) = \int_{t_0}^{t_1} f[t, x^*(t) + \varepsilon \eta(t), \dot{x}^*(t) + \varepsilon \dot{\eta}(t)] \, dt.$$

It is a consequence of (2.20) that

$$F(0) \le F(\varepsilon)$$

for all ε with $|\varepsilon|$ sufficiently small, and so the function $F(\cdot)$ has a local minimum at $\varepsilon = 0$ which is an interior point of its domain. Thus it is necessary that

$$F'(0) = 0. \qquad (2.21)$$

In what follows the arguments t, $x^*(t) + \varepsilon \eta(t)$, and $\dot{x}^*(t) + \varepsilon \dot{\eta}(t)$ are suppressed for the sake of brevity. Upon differentiating $F(\cdot)$, employing Leibniz's[†] rule (Ref. 2.2), we have

$$F'(\varepsilon) = \int_{t_0}^{t_1} (f_x \eta + f_{\dot{x}} \dot{\eta}) \, dt.$$

[†]G. W. Leibniz, German mathematician, 1646–1716.

Chap. 2 • Problem Statement

Setting $\varepsilon = 0$ and then integrating the first term by parts results in

$$\int_{t_0}^{t_1} f_x \eta \, dt = \eta(t) \int_{t_0}^{t} f_x \, d\tau \bigg|_{t_0}^{t_1} - \int_{t_0}^{t_1} \dot{\eta} \int_{t_0}^{t} f_x \, d\tau \, dt.$$

The first term on the right vanishes since $\eta(t_0) = \eta(t_1) = 0$, and so (2.21) becomes

$$\int_{t_0}^{t_1} \left[f_{\dot{x}} - \int_{t_0}^{t} f_x \, d\tau \right] \dot{\eta} \, dt = 0.$$

The function $m(\cdot)$ with values

$$m(t) = f_{\dot{x}}[t, x^*(t), \dot{x}^*(t)] - \int_{t_0}^{t} f_x[\tau, x^*(\tau), \dot{x}^*(\tau)] \, d\tau$$

is piecewise continuous on $[t_0, t_1]$. Thus we may employ it in the DuBois–Reymond Lemma 2.1 which leads at once to (2.19) and concludes the proof. □

The next theorem, which is a direct consequence of Theorem 2.1, gives another necessary condition.

Theorem 2.2. *If $x^*(\cdot)$ furnishes a weak local minimum of $J(\cdot)$ on $\mathcal{X}$, and if $t \in [t_0, t_1]$ is a point of continuity of $\dot{x}^*(\cdot)$, then*

$$\frac{d}{dt} f_{\dot{x}}[t, x^*(t), \dot{x}^*(t)] - f_x[t, x^*(t), \dot{x}^*(t)] = 0. \qquad (2.22)$$

Furthermore the so-called Euler–Lagrange[†] equation (2.22) applies as well at a point of discontinuity t of $\dot{x}^*(\cdot)$ for $\dot{x}^*(t-0)$ and $\dot{x}^*(t+0)$, respectively.

Proof. If $t \in (t_0, t_1)$ is a point of continuity of $\dot{x}^*(\cdot)$, then the integrand on the right of (2.19) is continuous at that point; then the relation (2.22) follows from the fundamental theorem of the calculus (Ref. 2.2). If $t \in (t_0, t_1)$ is a discontinuity point of $\dot{x}^*(\cdot)$, we obtain (2.22) at such a point for $\dot{x}^*(t-0)$ and $\dot{x}^*(t+0)$ by going to the limit from below and above, respectively. At t_0 and t_1, (2.22) is obtained similarly for $\dot{x}^*(t_0+0)$ and $\dot{x}^*(t_1-0)$, respectively. □

[†]J. L. Lagrange, French–Italian mathematician, 1736–1813.

Theorem 2.3. *If $x^*(\cdot)$ furnishes a weak local minimum of $J(\cdot)$ on $\mathfrak{X}$, if $t_2 \in [t_0, t_1]$ is a continuity point of $\dot{x}^*(\cdot)$, and if $f_{rr}[t_2, x^*(t_2), \dot{x}^*(t_2)] \neq 0$, then (i) $\ddot{x}^*(t_2)$ is defined, (ii) there is a subinterval I of $[t_0, t_1]$ containing t_2 such that $\ddot{x}^*(\cdot)$ is continuous on I, and (iii)*

$$f_x = f_{rt} + f_{rx}\dot{x}^*(t) + f_{rr}\ddot{x}^*(t) \qquad \forall t \in I \tag{2.23}$$

where f_x, f_{rt}, f_{rx} and f_{rr} have the arguments t, $x^*(t)$, and $\dot{x}^*(t)$. If the hypotheses hold for all points of $[t_0, t_1]$, so do the conclusions (i)–(iii).

Proof. Let $x_2^* \triangleq x^*(t_2)$ and $\dot{x}_2^* \triangleq \dot{x}^*(t_2)$. Also let $\Delta x^* \triangleq x^*(t_2 + \Delta t) - x_2^*$ and $\Delta \dot{x}^* \triangleq \dot{x}^*(t_2 + \Delta t) - \dot{x}_2^*$, where $\Delta t > 0$ or $\Delta t < 0$ for $t_2 \in (t_0, t_1)$, $\Delta t > 0$ for $t_2 = t_0$, and $\Delta t < 0$ for $t_2 = t_1$.

By the mean value theorem for the function $f_r(\cdot)$ of the three variables t, x, and r, there is a $\theta \in (0, 1)$ such that

$$f_r[t_2 + \Delta t, x_2^* + \Delta x^*, \dot{x}_2^* + \Delta \dot{x}^*] - f_r[t_2, x_2^*, \dot{x}_2^*]$$

$$\triangleq \Delta f_r = f_{rt}\Delta t + f_{rx}\Delta x^* + f_{rr}\Delta \dot{x}^* \tag{2.24}$$

where the arguments of f_{rt}, f_{rx}, and f_{rr} are $t_2 + \theta \Delta t$, $x_2^* + \theta \Delta x^*$ and $\dot{x}_2^* + \theta \Delta \dot{x}^*$; for instance, see Ref. 2.2.

The function $x^*(\cdot) \in \mathfrak{X}$ and so is piecewise smooth. By hypothesis, t_2 is a point of continuity of $\dot{x}^*(\cdot)$; hence, $\dot{x}^*(\cdot)$ is defined and continuous on a subinterval I' containing t_2. Consequently, $\Delta x^* \to 0$, $\Delta \dot{x}^* \to 0$ and $\Delta x^*/\Delta t \to \dot{x}^*(t_2)$ as $\Delta t \to 0$. Also, by Theorem 2.2, the left side of (2.24) divided by Δt possesses a limit equal to $f_x(t_2, x_2^*, \dot{x}_2^*)$.

By hypothesis, $f_{rr}(\cdot)$ is continuous and f_{rr} is nonzero at t_2; hence it is nonzero on a subinterval I'' containing t_2. Let $I = I' \cap I''$. Then it follows from (2.24) that $\Delta \dot{x}^*/\Delta t$ is a quotient with f_{rr} in the denominator. Each term in the numerator of that quotient has a finite limit; by hypothesis, the denominator f_{rr} has a nonzero limit. Thus, $\Delta \dot{x}^*/\Delta t$ has a finite limit which is $\ddot{x}^*(t_2)$ by definition, and (2.23) holds at t_2.

In similar fashion, one can show that $\ddot{x}^*(t)$ is defined—exists and is finite—for all $t \in I$; namely, $f_{rr}(\cdot)$ is continuous on $[t_0, t_1] \times R^2$, $\dot{x}^*(\cdot)$ is continuous on I', and $f_{rr}(t_2, x_2^*, \dot{x}_2^*) \neq 0$, so that there is a subinterval I'' containing t_2 such that $f_{rr}[t, x^*(t), \dot{x}^*(t)] \neq 0$ for $t \in I''$. Then, as before, $\ddot{x}^*(t)$ is defined on I. Thus, (2.23) holds on I, and so $\ddot{x}(\cdot)$ is continuous on I. $\square$

2.7. Remarks

Some comments concerning the results of Section 2.6 are in order.

(1) Theorems 2.1–2.3 are necessary conditions for a weak local minimum of $J(\cdot)$ on $\mathcal{X}$. Hence they are necessary for a strong local minimum as well as for a global minimum.

(2) Theorems 2.1–2.3 are also necessary for a *maximum* (of any kind) of $J(\cdot)$ on $\mathcal{X}$. This is so because they are based solely on the condition (2.21) which is necessary for *both* minima and maxima.

(3) Care must be taken in applying the necessary conditions embodied in Theorems 2.1–2.3. They hold under successively more stringent assumptions. Even in the least restricted one, Theorem 2.1, we assume in blanket fashion the existence of all required derivatives of $f(\cdot)$; thus the integrand $f(\cdot)$ must be such that $m(\cdot)$ is piecewise continuous on $[t_0, t_1]$. More restrictive assumptions are made in Theorems 2.2 and 2.3.

(4) Even if all hypotheses are met, all one can say is that if $x^*(\cdot)$ minimizes $J(\cdot)$ on $\mathcal{X}$ then the conditions of the relevant theorems must apply; that is, they are *necessary* only. Hence, a function $x^*(\cdot) \in \mathcal{X}$ satisfying (2.19) or (2.22) or (2.23) is only a *candidate* for a minimizing function. We say that it is an *extremal*; it need not furnish an extremum (minimum or maximum) of $J(\cdot)$. However, provided the appropriate hypotheses are met, *a minimizing function must be an extremal.*

(5) Usually the question asked is "Given $J(\cdot)$, find an $x^*(\cdot) \in \mathcal{X}$ such that $J(x^*(\cdot))$ is a minimum on $\mathcal{X}$." Then we can employ necessary conditions to obtain extremals. *If a minimizing function exists* and the hypotheses of the relevant theorem are met, then a minimizing function is one of the extremals. To verify that a particular extremal is indeed minimizing, one can proceed in two ways:

(i) Find *all* extremals. Prove the existence of a minimizing function in $\mathcal{X}$ and show that all hypotheses are met. Compare the values of $J(\cdot)$ for all extremals.
(ii) Employ *sufficient* conditions for a minimizing function to verify that a particular extremal is minimizing.

In any event, the usual first step is to utilize necessary conditions to deduce extremals. If Theorem 2.3 is used, that is, a solution of (2.23) is sought, then one deals with an ordinary second-order differential equation whose general solution involves two constants of integration, say α and β. Note that we are confronted with an *end value* problem rather than an *initial value* one, since the values of $x(t_0)$ and $x(t_1)$ rather than those of $x(t_0)$ and $\dot{x}(t_0)$ are specified. Thus, if we denote the general solution of (2.23) by $g(\cdot)\colon R^3 \to R^1$ where, given α and β,

$$x(t) = g(t, \alpha, \beta),$$

then it is required that

$$g(t_0, \alpha, \beta) = x_0, \qquad g(t_1, \alpha, \beta) = x_1.$$

It may happen that there is no $[\alpha \ \beta]^T \in R^2$ satisfying these end conditions. For instance, suppose that the first end condition can be solved for α as a function of β; that is, suppose there is a function $\hat{\alpha}(\cdot)\colon R^1 \to R^1$ such that

$$g(t_0, \hat{\alpha}(\beta), \beta) = x_0.$$

Then the function

$$g(\cdot, \hat{\alpha}(\beta), \beta)\colon [t_0, t_1] \to R^1, \qquad \beta \in R^1,$$

defines a one-parameter family of solutions of (2.23) passing through (t_0, x_0). It may happen that no member of this family of solutions passes through (t_1, x_1); this is illustrated in Figure 2.6.

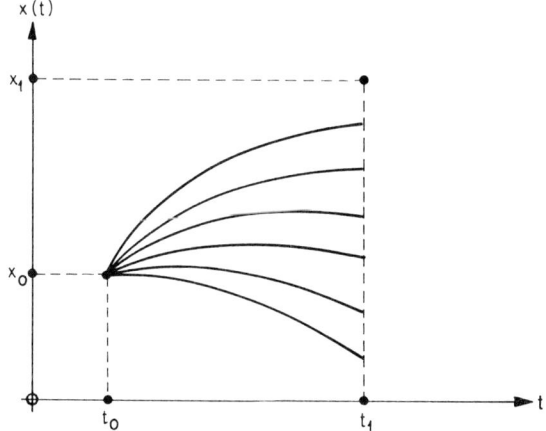

Figure 2.6. A family of extremals.

(6) Before proceeding we leave it as an exercise for the reader to deduce necessary conditions analogous to Theorems 2.1–2.3 for the case in which the integral $J(\cdot)$ depends on a *vector-valued* function. Consider

$$J(x(\cdot)) = \int_{t_0}^{t_1} f[t, x(t), \dot{x}(t)]\, dt$$

where now $f(\cdot):[t_0, t_1] \times R^{2n} \to R^1$, and $x(\cdot):[t_0, t_1] \to R^n$ is piecewise smooth such that $x(t_0) = x^0$ and $x(t_1) = x^1$ with x^0 and x^1 prescribed. Note the slight change in notation. We utilize superscripts 0 and 1 to denote specific values of the vector x since subscripts are used to denote components. Now $\mathfrak{X}$ is the set of all such *vector-valued* functions $x(\cdot)$.

For this situation all definitions introduced heretofore are extended to account for $x(\cdot)$ being vector-valued. For example, the distance of order 0 is now

$$d_0[x(\cdot), \bar{x}(\cdot)] \triangleq \sup_{t \in [t_0, t_1]} \|x(t) - \bar{x}(t)\|$$

where $\|\ \|$ denotes the Euclidean norm. The definition of $d_1[x(\cdot), \bar{x}(\cdot)]$ is altered in an analogous fashion.

Theorems 2.1–2.3 can then be readily extended by proving the relevant conditions for each component, $x_i^*(\cdot)$, of $x^*(\cdot)$. For instance, Theorem 2.1 has as its counterpart the following theorem.

Theorem 2.1'. *If $x^*(\cdot)$ furnishes a weak local minimum of $J(\cdot)$ on $\mathfrak{X}$, then there is a constant vector $c \triangleq [c_1\ c_2\ \cdots\ c_n]^T \in R^n$ such that for $i \in \{1, 2, \ldots, n\}$*

$$f_{x_i}[t, x^*(t), \dot{x}^*(t)] = \int_{t_0}^{t} f_{x_i}[\tau, x^*(\tau), \dot{x}^*(\tau)]\, d\tau + c_i$$

for all $t \in [t_0, t_1]$.

Proof. The proof goes essentially unaltered, except that now $\eta(t) \in R^n$. In applying Lemma 2.1 one takes $\eta_i(t) \neq 0$ and $\eta_j(t) \equiv 0$ for $j \neq i$. □

For the counterparts of Theorems 2.2 and 2.3 one proceeds as before, but now starting from Theorem 2.1'. Thus, for instance, in Theorem 2.2' corresponding to Theorem 2.2 there are now n Euler–Lagrange equations, one for each component of $x^*(\cdot)$.

2.8. Stationarity

Consider two admissible functions $x(\cdot) \in \mathcal{X}$ and $x(\cdot) + \varepsilon\eta(\cdot) \in \mathcal{X}$, $\varepsilon = $ constant. Then the difference in the corresponding values of the integral $J(\cdot)$, the so-called *total variation*, is

$$\Delta J \stackrel{\Delta}{=} J(x(\cdot) + \varepsilon\eta(\cdot)) - J(x(\cdot))$$
$$= F(\varepsilon) - F(0)$$
$$= \varepsilon F'(0) + o(\varepsilon), \qquad (2.25)$$

where

$$\lim_{\varepsilon \to 0} \frac{o(\varepsilon)}{\varepsilon} = 0.$$

The term of first order in ε,

$$\delta J \stackrel{\Delta}{=} \varepsilon F'(0), \qquad (2.26)$$

is called the *first variation* of $J(x(\cdot))$.

In Section 2.6 we showed that the vanishing of the first variation is necessary for $x(\cdot) = x^*(\cdot)$, if the function $x^*(\cdot)$ furnishes an extremum (minimum or maximum) of the integral $J(\cdot)$. However, here $x(\cdot)$ need not extremize $J(\cdot)$. Rather we consider two questions. The first is as follows:

(i) What condition must $x(\cdot)$ satisfy in order to result in $\delta J = 0$?

Utilizing the reasoning employed in Section 2.6, we see that $x(\cdot)$ must be an extremal; that is, at a continuity point of $\dot{x}(\cdot)$,

$$f_x[t, x(t), \dot{x}(t)] - \frac{d}{dt} f_{\dot{r}}[t, x(t), \dot{x}(t)] = 0 \qquad (2.27)$$

is a *necessary* condition for the vanishing of the first variation.

The second as yet unanswered question is the following:

(ii) If $x(\cdot)$ is an extremal, does δJ vanish?

To answer this question, consider

$$F'(0) = \int_{t_0}^{t_1} (f_x \eta + f_{\dot{r}} \dot{\eta}) \, dt$$

where $t, x(t), \dot{x}(t)$ are the arguments of f_x and $f_{\dot{r}}$. Now, unlike in the proof of Theorem 2.1 where the first term on the right is integrated by parts, let us

Chap. 2 • Problem Statement

integrate the second term by parts; that is,

$$\int_{t_0}^{t_1} f_{\dot{r}} \dot{\eta}\, dt = f_{\dot{r}} \eta(t) \Big|_{t_0}^{t_1} - \int_{t_0}^{t_1} \left(\frac{d}{dt} f_{\dot{r}} \right) \eta\, dt.$$

However, $\eta(t_0) = \eta(t_1) = 0$ so that

$$F'(0) = \int_{t_0}^{t_1} \left(f_x - \frac{d}{dt} f_{\dot{r}} \right) \eta\, dt. \tag{2.28}$$

Before proceeding we must note, however, that the integration by parts employed above in deducing (2.28) requires that $x(\cdot)$ is twice continuously differentiable, that is, is of class C^2; as in Theorem 2.3, this is so provided $\dot{x}(\cdot)$ is continuous and $f_{\dot{r}\dot{r}}[t, x(t), \dot{x}(t)] \neq 0$ on $[t_0, t_1]$.

Now it follows at once from (2.28) that $\delta J = 0$ if $x(\cdot)$ is an extremal, namely, a solution of the Euler–Lagrange equation (2.27). And so we can state the following lemma.

Lemma 2.2. *Given* $x(\cdot) \in \mathcal{X}$ *of class* C^2, *the first variation* δJ *of the integral* $J(x(\cdot))$ *vanishes if and only if* $x(\cdot)$ *is an extremal.*

The vanishing of the first variation is referred to as a *stationarity condition*, and the value $J(x(\cdot))$ furnished by an *extremal* $x(\cdot)$ is called a *stationary value* of $J(\cdot)$.

For some purposes, as in certain scientific principles such as Hamilton's principle in mechanics (for instance, see Refs. 2.4 and 2.5, also Section 4.3) one is concerned with stationarity rather than minimality of an integral.

Exercises

2.1. Consider the function $\phi(\cdot): [1,4] \to R^1$ with values

$$\phi(x) = x \quad \text{for } x \in [1,2],$$
$$\phi(x) = -x + 4 \quad \text{for } x \in [2,4].$$

Determine
(a) the local minimum (minima),
(b) the global minimum.

2.2. Consider the function $\phi(\cdot): [0, 4\pi] \to R^1$ with values $\phi(x) = \sin x$.
Determine
(a) the local minimum (minima),
(b) the global minimum.

2.3. Consider the function $\phi(\cdot):[0,4\pi]\to R^1$ with values $\phi(x)=-\sin x$. Determine
   (a) the local minimum (minima),
   (b) the global minimum.
   Show that the minima of $\phi(\cdot)$ are the maxima of $-\phi(\cdot)$.

2.4. Consider the function in Exercise 2. Apply the conditions (2.5) and (2.6).

2.5. Consider the function $\phi(\cdot):[0,4]\to R^1$ with values $\phi(x)=x^3-6x^2+9x+1$. Determine
   (a) the local minimum (minima),
   (b) the global minimum.
   Apply the conditions (2.5) and (2.6).

2.6. Consider the function $\phi(\cdot):[-1,4]\to R^1$ with $\phi(x)=3x^4-12x^3+12x^2-4$. Determine
   (a) the local minimum (minima),
   (b) the global minimum.
   Apply the conditions (2.5) and (2.6).

2.7. Consider the function $\phi(\cdot):[0,2]\to R^1$ with values $\phi(x)=x^2-4x+2$. Determine
   (a) the local minimum (minima),
   (b) the global minimum.
   Apply the conditions (2.5)–(2.7).

2.8. Given the two functions $x(\cdot)$ and $\bar{x}(\cdot)$ on $[0,2]$ with values $x(t)=2t$ and $\bar{x}(t)=t^3$, respectively, find $d_0[x(\cdot),\bar{x}(\cdot)]$ and $d_1[x(\cdot),\bar{x}(\cdot)]$.

2.9. Given the two functions $x(\cdot)$ and $\bar{x}(\cdot)$ on $[-1,1]$ with values $x(t)=2t$ and $\bar{x}(t)=2$, respectively, find $d_0[x(\cdot),\bar{x}(\cdot)]$ and $d_1[x(\cdot),\bar{x}(\cdot)]$.

2.10. Given the function $x(\cdot):[0,1]\to R^1$ with $x(t)=t$, find all linear functions $\bar{x}(\cdot):[0,1]\to R^1$, that is, $\bar{x}(t)=at+b$, a and $b\in R^1$, such that
   (a) $\bar{x}(\cdot)\in N_0[\delta,x(\cdot)]$ for $\delta=0.1$,
   and
   (b) $\bar{x}(\cdot)\in N_1[\delta,x(\cdot)]$ for $\delta=0.1$.

2.11. Utilize Theorems 2.1–2.3 to deduce necessary conditions for extrema of $J(\cdot)$ for
   (a) $f(t,x,r)=[1-r^2]^2$,
   (b) $f(t,x,r)=1-r^2$,
   (c) $f(t,x,r)=r^2-x^2$,
   (d) $f(t,x,r)=r^2-2xr-x^2$,
   (e) $f(t,x,r)=r^2+t^2$,
   (f) $f(t,x,r)=(1+r^2)^{1/2}$,
   (g) $f(t,x,r)=\sin(tr)$.

2.12. Discuss the applicability of Theorems 2.1 and 2.2 to the problem with $f(t,x,r)=[x(1-r^2)]^{1/2}$.

2.13. Using the alternative partial integration, show that $x(\cdot)$ in Lemma 2.2 need not be restricted to the class of C^2 functions.

3

Integration of the Euler–Lagrange Equation

3.1. Introduction

Let us suppose that the hypotheses of Theorem 2.3 are met so that the Euler–Lagrange equation (2.22) is of the form (2.23). Let us recall these hypotheses: (i) $f(\cdot)$ is twice continuously differentiable (of class C^2); (ii) $f_{rr}[t, x^*(t), \dot{x}^*(t)] \neq 0$ for all $t \in [t_0, t_1]$; (iii) $\ddot{x}^*(t)$ is defined for all $t \in [t_0, t_1]$. Then the Euler–Lagrange equation is

$$f_{rr}\ddot{x}^*(t) + f_{rx}\dot{x}^*(t) + f_{rt} - f_x = 0 \qquad \forall t \in [t_0, t_1] \tag{3.1}$$

where $t, x^*(t), \dot{x}^*(t)$ are the arguments of $f_{rr}, f_{rx}, f_{rt},$ and f_x.

To obtain extremals one must integrate (3.1). Under certain conditions this task is greatly simplified. We shall now consider some of these conditions and their consequences.

3.2. The Function $f(\cdot)$ Independent of x: $f_x(t, x, r) \equiv 0$

Since (3.1) is the expanded form of (2.22), namely, of

$$\frac{d}{dt}f_r - f_x = 0, \tag{3.2}$$

and since (3.2) is certainly valid under the hypotheses which permit (3.1), it follows at once that (3.1) possesses a *first integral*

$$f_r[t, x^*(t), \dot{x}^*(t)] = \text{constant} \tag{3.3}$$

on $[t_0, t_1]$.

3.3. The Function $f(\cdot)$ Independent of t: $f_t(t, x, r) \equiv 0$

Consider the derivative

$$\frac{d}{dt}[f - \dot{x}^*(t) f_r]$$

where t, $x^*(t)$, and $\dot{x}^*(t)$ are the arguments of f and f_r. Under our hypotheses we may expand this derivative; namely, in view of (2.22), we obtain

$$\frac{d}{dt}[f - \dot{x}^*(t) f_r] = f_t + \dot{x}^*(t) f_x + \ddot{x}^*(t) f_r - \ddot{x}^*(t) f_r - \dot{x}^*(t) f_{rt}$$
$$- [\dot{x}^*(t)]^2 f_{rx} - \dot{x}^*(t) \ddot{x}^*(t) f_{rr}$$
$$= \dot{x}^*(t) \left[f_x - \frac{d}{dt} f_r \right] = 0 \tag{3.4}$$

where the arguments t, $x^*(t)$, and $\dot{x}^*(t)$ have been omitted again for the sake of brevity.

Thus, if (2.22) holds, then

$$f - \dot{x}^*(t) f_r = \text{constant} \tag{3.5}$$

on $[t_0, t_1]$. In other words, every solution of the Euler–Lagrange equation satisfies (3.5). Conversely, every solution of (3.5), *except* possibly one for which $x^*(t) = \text{constant}$ on a subinterval of $[t_0, t_1]$, is a solution of the Euler–Lagrange equation; a solution of (3.5) with $x^*(t) \equiv \text{constant}$ is excepted since (3.5) follows from (3.4) if $\dot{x}^*(t) = 0$, even if (2.22) is not satisfied.

3.4. The Function $f(\cdot)$ Independent of t and x: $f_t(t, x, r) \equiv 0$ and $f_x(t, x, r) \equiv 0$

For this case it follows from (3.1) that

$$\ddot{x}^*(t) f_{rr} = 0 \quad \forall t \in [t_0, t_1].$$

Since $f_{rr} \neq 0$ by hypothesis, one must have

$$\ddot{x}^*(t) = 0 \quad \forall t \in [t_0, t_1]$$

and the extremals are straight lines, that is,

$$x^*(t) = a + bt, \tag{3.6}$$

where a and b are constants.

Finally it should be noted that hypothesis (iii) may be dropped; that is, if $x^*(\cdot)$ is only piecewise smooth so that $\dot{x}^*(\cdot)$ may have a finite number of discontinuities, then (3.3), (3.5), and (3.6) are still valid, however possibly with different constants in (3.6) for each subinterval on which $\dot{x}^*(\cdot)$ is continuous.

3.5. Examples

Example 3.1. Consider the integral

$$\int_{\xi_0}^{\xi_1} \phi\left[\left(\xi^2 + \eta^2(\xi)\right)^{1/2}\right] \left\{1 + [\eta'(\xi)]^2\right\}^{1/2} d\xi, \quad \eta(\xi_0) = \eta_0, \quad \eta(\xi_1) = \eta_1,$$

where $\phi(\cdot)$ is a given function and $\eta' \triangleq d\eta/d\xi$. Such an integral arises in the problem of determining minimum cost routes in an urban region (Ref. 3.1). On introducing polar coordinates (t, x), where

$$\xi = t \sin x, \quad \eta = t \cos x,$$

the integral becomes

$$\int_{t_0}^{t_1} \phi(t) \left\{1 + [t\dot{x}(t)]^2\right\}^{1/2} dt, \quad x(t_0) = x_0, \quad x(t_1) = x_1.$$

This integrand is of the type discussed in Section 3.2; hence the Euler-Lagrange equation has a first integral (3.3). Here we have

$$f(t, x, r) = \phi(t)[1 + (tr)^2]^{1/2}$$

so that (3.3) becomes

$$\phi(t) t^2 \dot{x}^*(t) \{1 + [t\dot{x}^*(t)]^2\}^{-1/2} = \text{constant} \triangleq c$$

whence, provided $t\phi(t) \neq 0$,

$$\dot{x}^*(t) = \pm(c/t)[t^2\phi^2(t) - c^2]^{-1/2}$$

and

$$x^*(t) = \pm \int_{t_0}^{t} \frac{c}{\tau} [\tau^2 \phi^2(\tau) - c^2]^{-1/2} d\tau + x_0.$$

Example 3.2. Consider the integrand

$$f(t, x, r) = (1 + r^2)^{1/2}/x$$

for $x \neq 0$ ($x > 0$ or $x < 0$). The exclusion of $x = 0$ is imposed in order to assure the finiteness of the integrand. For definiteness, suppose we impose $x > 0$. Thus far we have not admitted such *constraints*. We shall return to a discussion of constraints; for the moment let us proceed as if there were no constraint. However, let us impose end conditions which are consistent with it; namely, let

$$x(t_0) = x_0 > 0, \qquad x(t_1) = x_1 > 0.$$

The integrand is of the type discussed in Section 3.3. Thus, if $x^*(\cdot)$ is an extremal, then it must satisfy (3.5); that is,

$$f - \dot{x}^*(t) f_r = \left[x^*(t) \{1 + [\dot{x}^*(t)]^2\}^{1/2} \right]^{-1} = \text{constant} \triangleq \beta^{-1}$$

so that

$$x^*(t) = [\beta^2 - (t - \alpha)^2]^{1/2}, \qquad \alpha = \text{constant}.$$

Since $\dot{x}^*(t) \neq 0$ on a subinterval of $[t_0, t_1]$, every solution of (3.5) is an extremal. Thus the two-parameter family of extremals consists of circular arcs with the center on the t-axis; see Figure 3.1.

Example 3.3. Consider the integrand

$$f(t, x, r) = r^2$$

Chap. 3 • Integration of the Euler–Lagrange Equation

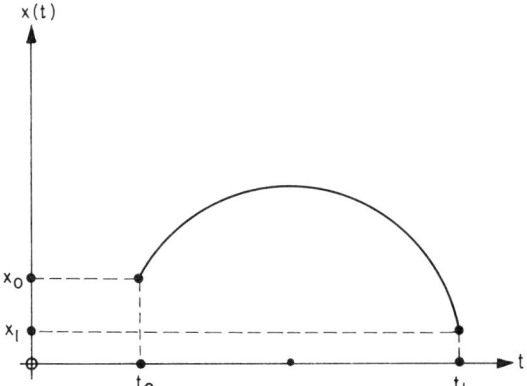

Figure 3.1. An extremal in Example 3.2.

and the end conditions

$$(t_0, x_0) = (0,0), \qquad (t_1, x_1) = (1,1).$$

Since the integrand is independent of t and x, it is of the kind discussed in Section 3.4. Thus the extremals are straight lines; that is,

$$x^*(t) = a + bt.$$

In view of the imposed end conditions, the particular extremal is given by $x^*(t) = t$.

Recall now that extremals are candidates for minimizing or maximizing functions. Here it is readily shown that there is no globally maximizing function. Consider the admissible function $x(\cdot) \in \mathcal{X}$ with the values

$$x(t) = \alpha t^2 + (1-\alpha)t, \qquad \alpha = \text{constant}.$$

Then

$$J(x(\cdot)) = \int_0^1 (2\alpha t + 1 - \alpha)^2 \, dt$$
$$= \tfrac{1}{3}\alpha^2 + 1.$$

But, as is readily seen,

$$J(x(\cdot)) \to \infty \qquad \text{as } \alpha \to \pm\infty,$$

which means that one can make $J(x(\cdot))$ arbitrarily large. Consequently there does not exist a *global maximum*. Indeed we shall show later that this integral does not even possess a strong or weak *local maximum*.

Example 3.4. The following example serves to illustrate a dictum ascribed to Professor Magnus Hestenes: "There is no discipline in which more correct results can be obtained by incorrect means than in the calculus of variations."

Consider the integrand

$$f(t, x, r) = \sqrt{r}$$

and the end conditions

$$(t_0, x_0) = (0,0), \qquad (t_1, x_1) = (1,0).$$

Here we have

$$f_r = \tfrac{1}{2} r^{-1/2}, \qquad f_{rr} = -\tfrac{1}{4} r^{-3/2}.$$

Thus not even Theorem 2.1 is applicable, since it is based on the assumption that $f(\cdot), f_r(\cdot)$, and $f_x(\cdot)$ are defined and continuous on $[t_0, t_1] \times R^2$. However, here $f(\cdot)$ is not defined for $r < 0$ and $f_r(\cdot)$ is not continuous at $r = 0$. Nonetheless, were we to apply (2.19) *blindly*, we would have

$$\tfrac{1}{2} [\dot{x}^*(t)]^{-1/2} = c$$

whence

$$[\dot{x}^*(t)]^{1/2} = 2/c$$

and

$$\dot{x}^*(t) \equiv 0 \qquad \text{for } c \to \infty.$$

Then, in view of the presribed end conditions,

$$x^*(t) \equiv 0.$$

Clearly, this furnishes a global minimum.

Example 3.5. Let us recall Dido's problem. In terms of Cartesian coordinates ξ and x, it is desired to maximize the area A enclosed between the ξ-axis and a curve C of given length l, starting at $\xi = x = 0$, ending at $x = 0$ with ξ unspecified, and lying in the upper half plane ($x \geq 0$); see Figure 3.2. In other words, we wish to determine the shape of the curve C of length l resulting in the maximum of area A. In terms of the variables ξ and x, this is not a problem of the kind considered thus far, since ξ_1 is not specified *and* since the class of admissible functions is restricted to those satisfying an integral constraint, a so-called *isoperimetric constraint*. However, if the variable ξ is dropped in favor of the arc length t, we can express the enclosed area by

$$A = -J(x(\cdot)) = \int_0^l x(t) \left[1 - \dot{x}^2(t) \right]^{1/2} dt$$

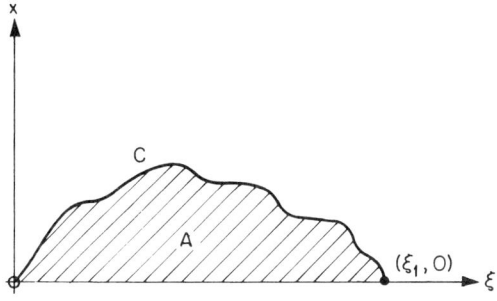

Figure 3.2. Dido's problem.

subject to the end conditions

$$x(0)=0, \quad x(l)=0.$$

The problem of maximizing A on $\mathfrak{X}$ is now of the kind discussed earlier. Note that this choice of variables allows for the trivial satisfaction of the isoperimetric constraint

$$\int_0^l dt = l.$$

However, there are two other constraints, one *imposed*, namely,

$$x(t) \geq 0 \quad \forall t \in [0, l]$$

and the other *implicit* to assure the existence of the integrand, namely,

$$|\dot{x}(t)| \leq 1 \quad \forall t \in [0, l].$$

We shall not pursue the solution to this problem, but rather leave it as an exercise for the reader once we have discussed the ramifications of the two constraints; see Exercise 3.2, also Sections 13.12 and 13.13, and Exercise 15.8.

3.6. Remarks

As we saw in Example 3.5, there may arise two kinds of constraints on $x(t)$ and $\dot{x}(t)$, *a priori* imposed ones and implicit ones; the latter are due to mathematical requirements such as existence. At this stage of the development we do not admit any constraints on $x(t)$ and $\dot{x}(t)$. How shall we proceed?

First of all, let us remark that the continuity properties imposed on $f(\cdot)$ and its partial derivatives in Theorems 2.1–2.3 need not be *global*; that is, they need not hold on $[t_0, t_1] \times R^2$. The validity of these theorems is unaffected if continuity properties hold *locally*, that is, for all $t \in [t_0, t_1]$ and all $[x \ r]^T$ in a neighborhood of $[x^*(t) \ \dot{x}^*(t)]^T$. Thus, provided $[x^*(t) \ \dot{x}^*(t)]^T$ does not belong to the boundary of the constraint set—for example, if $x^*(t) > 0$ and $|\dot{x}^*(t)| < 1$ for all $t \in [t_0, t_1]$—such a neighborhood exists and the theorems remain valid.

It may happen that $[x^*(t) \ \dot{x}^*(t)]^T$ belongs to the interior of the constraint set except at $t = t_0$ and $t = t_1$. In that case we proceed as follows.

Suppose $x^*(\cdot) \in \mathcal{X}$ is a minimizing function for $J(\cdot)$ on $\mathcal{X}$. If local minima are under discussion, we replace $\mathcal{X}$ by $\mathcal{X} \cap N_1[\delta, x^*(\cdot)]$ or $\mathcal{X} \cap N_0[\delta, x^*(\cdot)]$. Let t_2 and t_3, respectively, be interior points of the first and last continuity subinterval of $\dot{x}^*(\cdot)$; that is, t_2 and t_3 are such that $\dot{x}^*(\cdot)$ is continuous on $[t_0, t_4]$ for some t_4, $t_4 > t_2 > t_0$, and on $[t_5, t_1]$ for some t_5, $t_5 < t_3 < t_1$. Now consider all $x(\cdot) \in \mathcal{X}$ such that

$$x(t) = x^*(t) \quad \forall t \in [t_0, t_2] \cup [t_3, t_1].$$

Denote this subset by $\mathcal{X}'$. For an illustration see Figure 3.3. Then it follows from

$$J(x^*(\cdot)) \leq J(x(\cdot)) \quad \forall x(\cdot) \in \mathcal{X}$$

that

$$\int_{t_2}^{t_3} f[t, x^*(t), \dot{x}^*(t)] \, dt \leq \int_{t_2}^{t_3} f[t, x(t), \dot{x}(t)] \, dt$$

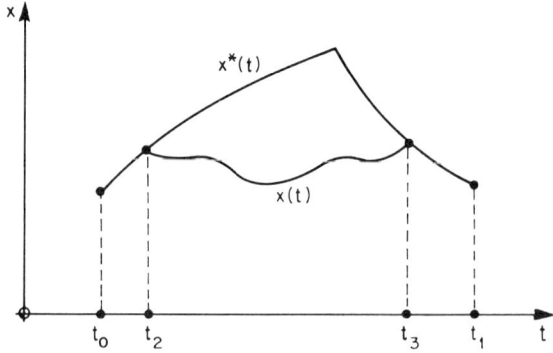

Figure 3.3. A function $x(\cdot) \in \mathcal{X}'$.

Chap. 3 • Integration of the Euler–Lagrange Equation

for all $x(\cdot)\in\mathcal{X}'$. Thus the restriction of $x^*(\cdot)$ to a subinterval $[t_2, t_3]$ is a minimizing function for $\int_{t_2}^{t_3} f[t, x(t), \dot{x}(t)] \, dt$ with respect to all piecewise smooth $x(\cdot):[t_2, t_3]\to R^1$ such that $x(t_2)=x^*(t_2), x(t_3)=x^*(t_3)$. Consequently Theorems 2.1–2.3 apply with $t\in[t_2, t_3]$, and since $x^*(\cdot)$ and $\dot{x}^*(\cdot)$ are continuous on $[t_0, t_4]$ and $[t_5, t_1]$, we may let $t_2\to t_0$ and $t_3\to t_1$.

The reader should convince himself that the reasoning employed above is valid in Dido's problem, Example 3.5. On the other hand, in Example 3.4 the constraint is $r\ge 0$ and $\dot{x}^*(t)\equiv 0$.

3.7. The Function $f(\cdot)$ Linear in r: $f_{rr}(t, x, r)\equiv 0$

Another interesting case is the one in which r enters $f(\cdot)$ linearly. Two possibilities must be considered: The Euler–Lagrange equation (2.22) is a finite equation or it is an identity.

If

$$f_{rr}(t, x, r)=0 \qquad \forall [t \; x \; r]^T\in[t_0, t_1]\times R^2 \qquad (3.7)$$

then

$$f(t, x, r)=M(t, x)+N(t, x)r \qquad (3.8)$$

where $M(\cdot)$ and $N(\cdot):[t_0, t_1]\times R^1\to R^1$ are differentiable functions. On substituting (3.8) in (2.22), one has

$$M_x[t, x^*(t)]-N_t[t, x^*(t)]=0. \qquad (3.9)$$

If (3.9) is *not* an identity—that is, if it is not satisfied for all $[t \; x^*(t)]^T \in [t_0, t_1]\times R^1$—then it is a finite equation; that is, $(t, x^*(t))$ is a point on a curve whose equation is (3.9). In general, this curve does not pass through particular given points (t_0, x_0) and (t_1, x_1). In other words, equation (3.9) furnishes a *single* extremal.

If (3.9) is an identity, namely, if

$$M_x(t, x)=N_t(t, x) \qquad \forall [t \; x]^T\in[t_0, t_1]\times R^1, \qquad (3.10)$$

and if $M(\cdot)$ and $N(\cdot)$ are functions of class C^1, then there exists a function $V(\cdot):[t_0, t_1]\times R^1\to R^1$ of class C^1 such that

$$V_t(t, x)=M(t, x), \qquad V_x(t, x)=N(t, x) \qquad (3.11)$$

so that

$$f(t, x, r) = V_t(t, x) + V_x(t, x)r. \qquad (3.12)$$

This follows (Ref. 3.2) from the fact that (3.10) is a necessary and sufficient condition for the integrability of

$$M(t, x)\, dt + N(t, x)\, dx.$$

Here we need concern ourselves only with the sufficient condition: If (3.10) holds, then there exists a C^1 function $V(\cdot)$ such that

$$M(t, x)\, dt + N(t, x)\, dx = dV(t, x). \qquad (3.13)$$

To prove this assertion we proceed as follows. Let

$$F(t, x) \stackrel{\Delta}{=} \int M(t, x)\, dt$$

so that

$$M(t, x) = F_t(t, x)$$

and then, in view of (3.10),

$$F_{tx}(t, x) = F_{xt}(t, x) = M_x(t, x) = N_t(t, x).$$

Thus one has

$$\frac{\partial}{\partial t}[N(t, x) - F_x(t, x)] = 0$$

so that there exists a continuous function $\Phi(\cdot): R^1 \to R^1$ such that

$$N(t, x) - F_x(t, x) = \Phi(x).$$

Now let

$$V(t, x) \stackrel{\Delta}{=} F(t, x) + \int \Phi(x)\, dx.$$

Consequently we have

$$N(t, x) = F_x(t, x) + \Phi(x) = V_x(t, x)$$

Chap. 3 • Integration of the Euler–Lagrange Equation

and

$$M(t, x) = F_t(t, x) = V_t(t, x).$$

This concludes the proof of the assertion.

Now, returning to (3.12), we consider $x(\cdot) \in \mathcal{X}$ and

$$J(x(\cdot)) = \int_{t_0}^{t_1} f[t, x(t), \dot{x}(t)] \, dt$$

$$= \int_{t_0}^{t_1} [V_t + V_x \dot{x}(t)] \, dt,$$

where t and $x(t)$ are the arguments of V_t and V_x. It follows that

$$J(x(\cdot)) = V(t_1, x_1) - V(t_0, x_0). \tag{3.14}$$

In other words, if (3.7) holds *and* if the consequent Euler–Lagrange equation (3.9) is an identity, then

$$J(x(\cdot)) = \text{constant} \quad \forall x(\cdot) \in \mathcal{X},$$

that is, the value of $J(\cdot)$ is independent of the choice of $x(\cdot) \in \mathcal{X}$.

Now we are ready to prove the following theorem.

Theorem 3.1. *Given the function $f(\cdot)$, the value of the integral $J(\cdot)$,*

$$J(x(\cdot)) = \int_{t_0}^{t_1} f[t, x(t), \dot{x}(t)] \, dt,$$

is independent of the choice of $x(\cdot) \in \mathcal{X}$, if and only if

$$f_x[t, x(t), \dot{x}(t)] - \frac{d}{dt} f_{\dot{x}}[t, x(t), \dot{x}(t)] = 0 \tag{3.15}$$

for all $[t \ x(t) \ \dot{x}(t) \ \ddot{x}(t)]^T \in [t_0, t_1] \times R^3$, that is, the Euler–Lagrange equation is an identity.

Proof. First let us demonstrate the *sufficient* (if) part of the assertion. Note that $\ddot{x}(t)$ appears linearly in (3.15) and, as a consequence of the hypothesized arbitrariness of $[t \ x(t) \ \dot{x}(t) \ \ddot{x}(t)]^T$, the coefficient of $\ddot{x}(t)$ must vanish identically; that is, (3.7) must hold. But as we showed just before stating Theorem 3.1, if (3.7) holds and the Euler–Lagrange equation is an identity, then $J(x(\cdot))$ is independent of $x(\cdot) \in \mathfrak{X}$.

To prove the necessary (only if) portion of the assertion, we suppose that $J(x(\cdot)) = $ constant for all $x(\cdot) \in \mathfrak{X}$. So let us consider an $x(\cdot) \in \mathfrak{X}$ such that, given $t_2 \in (t_0, t_1)$, $x_2 \in R^1$ and constants α and β,

$$x(t_2) = x_2, \quad \dot{x}(t_2) = \alpha, \quad \ddot{x}(t_2) = \beta.$$

This can always be done; for example, $x(t) = \Sigma_{i=0}^4 a_i t^i$, where the a_i are appropriate constants. Also, as we did in deriving the Euler–Lagrange equation, consider again functions $x(\cdot) + \varepsilon \eta(\cdot) \in \mathfrak{X}$ and the corresponding value of $J(\cdot)$, namely,

$$F(\varepsilon) = J(x(\cdot) + \varepsilon \eta(\cdot)).$$

Then, by hypothesis,

$$F(\varepsilon) = F(0)$$

for all ε, so that

$$F'(\varepsilon) = 0$$

and, in particular,

$$F'(0) = 0.$$

Then, as in Theorem 2.2, the Euler–Lagrange equation (2.22) must be satisfied, but now identically on $(t_0, t_1) \times R^3$ and hence by continuity on $[t_0, t_1] \times R^3$. □

Example 3.6. Consider the integrand $f(t, x, r) = 3t^2 x^2 + 2t^3 xr$. Then

$$f_x[t, x(t), \dot{x}(t)] = 6t^2 x(t) + 2t^3 \dot{x}(t),$$
$$f_r[t, x(t), \dot{x}(t)] = 2t^3 x(t)$$

whence

$$\frac{d}{dt} f_r[t, x(t), \dot{x}(t)] \equiv f_x[t, x(t), \dot{x}(t)],$$

that is, the Euler–Lagrange equation is an identity.

Chap. 3 • Integration of the Euler–Lagrange Equation

Furthermore,
$$V_t(t,x) = 3t^2 x^2, \qquad V_x(t,x) = 2t^3 x$$

so that
$$dV(t,x) = 3t^2 x^2 \, dt + 2t^3 x \, dx$$

and hence
$$V(t,x) = t^3 x^2 + \text{constant}.$$

Thus, given any $x(\cdot) \in \mathcal{X}$ and the end conditions $x(t_0) = x_0$, $x(t_1) = x_1$, we arrive at
$$J(x(\cdot)) = V(t_1, x_1) - V(t_0, x_0)$$
$$= t_1^3 x_1^2 - t_0^3 x_0^2.$$

Exercises

3.1. Give the general solution (family of extremals) of the Euler–Lagrange equation corresponding to each of the integrals with the integrands listed in Exercise 2.11.

3.2. Deduce extremals for Dido's problem, Example 3.5, and discuss the applicability of Theorems 2.1–2.3.

3.3. Consider the minimal cost routing problem, Example 3.1. Given $\phi(t) = \alpha t^{-1}$, $\alpha = \text{const} > 0$, deduce the corresponding extremals.

3.4. Consider a particle of unit mass required to move in rectilinear motion from the position $x(t_0) = x_0$ to $x(t_1) = x_1$ in the prescribed time $t_1 - t_0$. Determine the extremal for the problem of minimizing the particle's average kinetic energy
$$(t_1 - t_0)^{-1} \int_{t_0}^{t_1} \frac{1}{2} \dot{x}^2(t) \, dt.$$

3.5. Deduce the extremal(s) for the integral with integrand $f(t, x, r) = t^2 + x^2 + xr$.

3.6. Recall the problem of range maximization for a rocket plane in horizontal flight, posed in Chapter 1. Deduce the extremal(s) for this problem.

4

An Inverse Problem

4.1. Problem Statement

Heretofore we considered the problem of deducing necessary conditions for the minimality, and hence the stationarity, of a given integral $J(\cdot)$ on the set of admissible functions $\mathcal{X}$, that is, we prescribed the integrand function $f(\cdot)$ and sought the corresponding extremals. Before continuing with a discussion of other aspects of this problem let us make a detour and consider the following *inverse* problem: Find the set of integrand functions $f(\cdot)$ such that, given the function $g(\cdot): R^3 \to R^1$ of class C^2, and given $\alpha \in R^1$ and $\beta \in R^1$, the function $x(\cdot):[t_0, t_1] \to R^1$ with the values

$$x(t) = g(t, \alpha, \beta), \quad t \in [t_0, t_1], \tag{4.1}$$

is a solution of the Euler–Lagrange equation

$$f_{rr}\ddot{x}(t) + f_{rx}\dot{x}(t) + f_{rt} - f_x = 0, \tag{4.2}$$

where t, $x(t)$, $\dot{x}(t)$ are the arguments of f_{rr}, f_{rx}, f_{rt}, and f_x.

In other words, given a two-parameter family of functions, we inquire after the integrals which are made stationary by these functions; or to put it yet another way, we ask for the integrals for which (4.1) are the values of extremals.

4.2. Solution

Since we utilize the Euler–Lagrange equation in expanded form (4.2), we impose the restriction

$$f_{rr}(t, x, r) \neq 0 \quad \forall [t \ x \ r]^T \in [t_0, t_1] \times R^2, \tag{4.3}$$

in addition to the tacit one concerning the required differentiability of $f(\cdot)$.

In order to obtain the differential equation whose general solution is given by (4.1), we assume that one can eliminate the constants of integration α and β that is, we assume that there exist continuous functions $\phi(\cdot):[t_0,t_1]\times R^2 \to R^1$ and $\psi(\cdot):[t_0,t_1]\times R^2 \to R^1$ such that

$$x(t) = g[t, \phi(t, x(t), \dot{x}(t)), \psi(t, x(t), \dot{x}(t))],$$
$$\dot{x}(t) = g_t[t, \phi(t, x(t), \dot{x}(t)), \psi(t, x(t), \dot{x}(t))]. \quad (4.4)$$

In other words, speaking loosely, we assume that

$$x(t) = g(t, \alpha, \beta), \qquad \dot{x}(t) = g_t(t, \alpha, \beta)$$

can be solved for α and β. Then

$$\ddot{x}(t) = G[t, x(t), \dot{x}(t)] \quad (4.5)$$

where $G(\cdot):[t_0,t_1]\times R^2 \to R^1$ has the values

$$G(t, x, r) = g_{tt}[t, \phi(t, x, r), \psi(t, x, r)].$$

Equation (4.5) is the differential equation with the general solution furnished by (4.1). Now we seek $f(\cdot)$ such that (4.5) is the Euler-Lagrange equation (4.2), that is,

$$f_x - f_{rt} - \dot{x}(t) f_{rx} = G f_{rr}. \quad (4.6)$$

Since (4.1) gives the values of the general solution of (4.5), it must furnish a solution for every initial condition $[x(t_0)\ \dot{x}(t_0)]^T \in R^2$. Thus the relation

$$f_x - f_{rt} - r f_{rx} = G f_{rr}, \quad (4.7)$$

where t, x, r are the arguments of f_x, f_{rt}, f_{rx}, f_{rr}, and G, holds for all $[t\ x\ r]^T \in [t_0,t_1]\times R^2$. In particular, it is an identity in r and hence may be differentiated with respect to r. After letting

$$\mathcal{M}(t, x, r) \triangleq f_{rr}(t, x, r) \quad (4.8)$$

and assuming that

$$f_{xr} = f_{rx}, \qquad f_{rtr} = f_{rrt}, \qquad f_{rxr} = f_{rrx},$$

we obtain

$$\frac{\partial \mathcal{M}}{\partial t} + r\frac{\partial \mathcal{M}}{\partial x} + G\frac{\partial \mathcal{M}}{\partial r} + G_r\mathcal{M} = 0. \qquad (4.9)$$

This is a partial differential equation for the variable $\mathcal{M}(\cdot):[t_0, t_1] \times R^2 \to R^1$; for instance, see Ref. 4.1. The general solution of (4.9), as can be verified by substitution, is given by

$$\mathcal{M}(t, x, r) = \frac{\Phi[\phi(t, x, r), \psi(t, x, r)]}{\Theta[t, \phi(t, x, r), \psi(t, x, r)]} \qquad (4.10)$$

where

$$\Theta(t, \alpha, \beta) \triangleq \exp\left\{ \int G_r[t, g(t, \alpha, \beta), g_t(t, \alpha, \beta)] \, dt \right\} \qquad (4.11)$$

and $\Phi(\cdot): R^2 \to R^1$ is a differentiable, nonzero but otherwise arbitrary function.

The sought-after functions $f(\cdot)$ are then obtained from (4.10) by two successive quadratures; namely,

$$f(t, x, r) = \int_0^r \int_0^q \mathcal{M}(t, x, p) \, dp \, dq + r\lambda(t, x) + \mu(t, x) \qquad (4.12)$$

where $\lambda(\cdot):[t_0, t_1] \times R^1 \to R^1$ and $\mu(\cdot):[t_0, t_1] \times R^1 \to R^1$ are arbitrary except for the requirement that $f(\cdot)$ be such that (4.7) is satisfied.

In view of the arbitrariness of $\Phi(\cdot)$, $\lambda(\cdot)$, and $\mu(\cdot)$, we conclude that there exists an infinity of integrand functions $f(\cdot)$ for which (4.1) furnishes extremals.

4.3. Examples

Example 4.1. Determine integrand functions $f(\cdot)$ for which the extremals are straight lines, that is,

$$x(t) = \alpha t + \beta.$$

It follows from (4.4) that

$$\phi(t, x, r) = r, \qquad \psi(t, x, r) = x - tr.$$

Also, since $\ddot{x}(t) \equiv 0$,
$$G(t, x, r) \equiv 0,$$
and so, by (4.11),
$$\Theta(t, \alpha, \beta) = 1,$$
and then, by (4.10),
$$\mathfrak{M}(t, x, r) = \Phi(r, x - tr).$$

To obtain $f(\cdot)$, we integrate by parts as follows:
$$\int_0^r \int_0^q \mathfrak{M}(t, x, p)\, dp\, dq = q\int_0^q \mathfrak{M}(t, x, p)\, dp\bigg|_0^r - \int_0^r q\mathfrak{M}(t, x, q)\, dq$$
$$= \int_0^r (r-q)\mathfrak{M}(t, x, q)\, dq.$$

Thus, (4.12) becomes
$$f(t, x, r) = \int_0^r (r-q)\Phi(q, x - tq)\, dq + r\lambda(t, x) + \mu(t, x)$$

where $f(\cdot)$ must be such that (4.7) is satisfied. Let $z \stackrel{\triangle}{=} x - tq$ so that $\Phi(q, x - tq) = \Phi(q, z)$ and hence
$$\Phi_x = \Phi_z \frac{\partial z}{\partial x} = \Phi_z,$$
$$\Phi_t = \Phi_z \frac{\partial z}{\partial t} = -q\Phi_z.$$

Substitution in (4.7) then results in
$$\frac{\partial \lambda}{\partial t} = \frac{\partial \mu}{\partial x}$$

so that there is a function $\nu(\cdot): R^2 \to R^1$ such that
$$\lambda(t, x) = \frac{\partial \nu}{\partial x}, \qquad \mu(t, x) = \frac{\partial \nu}{\partial t},$$
which is the sole restriction on $\lambda(\cdot)$ and $\mu(\cdot)$.

Example 4.2. Consider a unit mass subjected to a force $P(t, x)$ where t is time and x is displacement. For rectilinear motion it follows from Newton's second law that
$$\ddot{x}(t) = P(t, x(t)).$$

Chap. 4 • An Inverse Problem

Suppose now that $P(t, x)$ is derivable from a potential; that is, there is a function $U(\cdot): R^2 \to R^1$ such that

$$P(t, x) = \frac{\partial U(t, x)}{\partial x}.$$

Then it follows from Hamilton's principle (for example, Refs. 4.2 and 4.3) that the equation of the particle's motion is the Euler–Lagrange equation for the integral

$$J(x(\cdot)) = \int_{t_0}^{t_1} \left[\frac{1}{2} \dot{x}^2(t) + U(t, x(t)) \right] dt$$

with $x(t_0) = x_0$, $x(t_1) = x_1$; namely,

$$\ddot{x}(t) - \frac{\partial U(t, x(t))}{\partial x} = 0.$$

Of course, this is merely Newton's second law again.

We are led to the question: Aside from Hamilton's principle, are there stationarity principles which apply even if the force $P(\cdot)$ is not derivable from a potential $U(\cdot)$, say a force that depends on t, $x(t)$, and $\dot{x}(t)$? We consider an equation of motion of the form

$$\ddot{x}(t) = G[t, x(t), \dot{x}(t)]. \qquad (4.13)$$

If the general solution of (4.13) were given, one could deduce an infinity of integrals for which (4.13) is the Euler–Lagrange equation. However, our problem is different; namely, we wish to determine the force function $G(\cdot)$ and a corresponding integral for which (4.13) is the Euler–Lagrange equation; see Ref. 4.4.

Let $a(\cdot): R^2 \to R^1$, $b(\cdot): R^1 \to R^1$ and $c(\cdot): R^1 \to R^1$ be functions of class C^1, and consider

$$G(t, x, r) = a(t, x) + b(t)r + c(x)r^2. \qquad (4.14)$$

Without specifying the functions $a(\cdot)$, $b(\cdot)$, and $c(\cdot)$, we cannot even attempt to solve (4.13) and to give its general solution. However, let us recall that the function $\Phi(\cdot)$ in (4.10) is arbitrary except for being differentiable and nonzero. Consider

$$\Phi(\alpha, \beta) \equiv 1$$

so that

$$\mathfrak{M}(t, x, r) = \Theta^{-1}[t, \phi(t, x, r), \psi(t, x, r)].$$

But by (4.11) with (4.14),

$$\Theta[t, \phi(t, x, r), \psi(t, x, r)] = \exp\left[\int b(t)\, dt + 2 \int c(x)\, dx \right] \stackrel{\triangle}{=} \theta(t, x) \qquad (4.15)$$

whence, by (4.12),

$$f(t, x, r) = \frac{1}{2}\theta^{-1}(t, x)r^2 + \int a(t, x)\theta^{-1}(t, x)\, dx. \qquad (4.16)$$

In other words, the equation

$$\ddot{x}(t) = a[t, x(t)] + b(t)\dot{x}(t) + c[x(t)]\dot{x}^2(t)$$

is the Euler–Lagrange equation for the integral

$$\int_{t_0}^{t_1} f[t, x(t), \dot{x}(t)]\, dt$$

with $f(\cdot)$ given by (4.16).

If the force function $G(\cdot)$ is derivable from a potential, that is, if

$$G(t, x, r) = a(t, x) = \frac{\partial U(t, x)}{\partial x},$$

then (4.16) becomes

$$f(t, x, r) = \tfrac{1}{2}r^2 + U(t, x)$$

and we arrive again at Hamilton's principle as a special case.

Exercises

4.1. Consider the two-parameter family of functions with values

$$x(t) = \left[\beta^2 - (t-\alpha)^2\right]^{1/2}$$

with α, β = constant. Find integrand functions $f(\cdot)$ such that the functions $x(\cdot)$ are extremals for

$$J(x(\cdot)) = \int_{t_0}^{t_1} f[t, x(t), \dot{x}(t)]\, dt.$$

4.2. Consider the equation of motion of an undamped linear oscillator

$$\ddot{x}(t) = -kx(t)$$

where k = constant > 0. Deduce a corresponding stationarity principle; that is, find a function $f(\cdot)$ such that $\delta J = 0$, where

$$J(x(\cdot)) = \int_{t_0}^{t_1} f[t, x(t), \dot{x}(t)]\, dt.$$

Chap. 4 • An Inverse Problem

4.3. Consider a damped linear oscillator whose equation of motion is

$$\ddot{x}(t) = -kx(t) - l\dot{x}(t)$$

where k and l are positive constants. Derive a corresponding stationarity principle.

4.4. Consider a system described by

$$\ddot{x}(t) = \omega[t, x(t)]\dot{x}^2(t), \qquad \dot{x}(t) > 0$$

where $\omega(\cdot)$ is a function of class C^1 on R^2. Show that there is a corresponding stationarity principle, $\delta J = 0$, with the integrand

$$f(t, x, r) = -\ln r + \int \omega(t, x)\, dx.$$

4.5. Consider a system described by

$$\ddot{x}(t) + k(t)\dot{x}(t) + l(t)m[x(t)] = 0$$

where $k(\cdot)$, $l(\cdot)$, and $m(\cdot)$ are functions of class C^1 on R^1. Show that there is a corresponding stationarity principle, $\delta J = 0$, with the integrand

$$f(t, x, r) = \left[\frac{1}{2}r^2 - l(t)\int m(x)\, dx\right]\exp\left[\int k(t)\, dt\right].$$

5

The Weierstrass Necessary Condition

5.1. Introduction

We return now to a consideration of further necessary conditions for a minimizing function $x^*(\cdot)$ of the integral $J(\cdot)$. Thus far we have shown the necessity of the Euler condition in its various forms; we recall that it is a necessary condition for both a weak local, and hence a strong local as well as a global minimum *and* maximum of $J(\cdot)$. Now we shall deduce a condition, the Weierstrass† condition, that is necessary for a strong local, and hence a global minimum of $J(\cdot)$; it is not necessary for a *weak* local minimum.

5.2. The Excess Function and a Necessary Condition

Let $E(\cdot):[t_0,t_1]\times R^3 \to R^1$ be the function with the values

$$E(t,x,r,q) \triangleq f(t,x,q) - f(t,x,r) - (q-r)f_r(t,x,r). \qquad (5.1)$$

This function, called the Weierstrass *excess function*, is utilized in the following theorem.

Theorem 5.1. *If* $x^*(\cdot) \in \mathfrak{X}$ *furnishes a strong local minimum of* $J(\cdot)$ *on* $\mathfrak{X}$, *then*

$$E[t, x^*(t), \dot{x}^*(t), q] \geq 0 \qquad (5.2)$$

for all $t \in [t_0, t_1]$ *and all* $q \in R^1$. *If* $\dot{x}^*(\cdot)$ *is discontinuous at* $t \in (t_0, t_1)$, *this condition applies for* $\dot{x}^*(t-0)$ *and* $\dot{x}^*(t+0)$.

†Karl Weierstrass, German mathematician, 1815–1879.

Proof. To prove this theorem let us recall the function $x(\cdot)$ used in Section 2.4 to exhibit a function belonging to a δ-neighborhood of order 0, $N_0[\delta, x^*(\cdot)]$. We require such a function since we are considering a *strong* local minimum. Let $[a, b] \subset [t_0, t_1]$ be an interval between successive corners of $x^*(\cdot)$, with $a \in [t_0, t_1)$. Let

$$x(t) = \begin{cases} x^*(t) & \text{for } t \in [t_0, a] \cup [b, t_1], \\ X(t) & \text{for } t \in [a, \varepsilon], \quad \varepsilon \in [a, b), \\ \phi(t, \varepsilon) & \text{for } t \in [\varepsilon, b] \end{cases}$$

where

$$X(t) \triangleq x^*(a) + q(t-a), \quad q \in R^1,$$

$$\phi(t, \varepsilon) \triangleq x^*(t) + \frac{X(\varepsilon) - x^*(\varepsilon)}{b - \varepsilon}(b - t).$$

The function $x(\cdot)$ is illustrated in Figure 2.5.

Next consider the function $\Phi(\cdot): [a, b) \to R^1$ with the values

$$\Phi(\varepsilon) \triangleq J(x(\cdot)) - J(x^*(\cdot));$$

that is, for given ε, $\Phi(\varepsilon)$ is the difference in the values of the integral $J(\cdot)$ furnished by $x(\cdot)$ and $x^*(\cdot)$ respectively.

Since $J(x^*(\cdot)) \leq J(x(\cdot))$ by hypothesis,

$$\Phi(\varepsilon) \geq 0 \quad \text{for } \varepsilon \in [a, b),$$

and

$$\Phi(a) = 0.$$

Thus it is necessary that

$$\Phi'(a) \geq 0. \tag{5.3}$$

To employ this condition we require

$$\Phi(\varepsilon) = \int_a^\varepsilon \left\{ f[t, X(t), \dot{X}(t)] - f[t, x^*(t), \dot{x}^*(t)] \right\} dt$$

$$+ \int_\varepsilon^b \left\{ f[t, \phi(t, \varepsilon), \phi_t(t, \varepsilon)] - f[t, x^*(t), \dot{x}^*(t)] \right\} dt$$

Chap. 5 • The Weierstrass Necessary Condition

so that, on using Leibniz's rule, we obtain

$$\Phi'(\varepsilon) = f[\varepsilon, X(\varepsilon), \dot{X}(\varepsilon)] - f[\varepsilon, \phi(\varepsilon, \varepsilon), \phi_t(\varepsilon, \varepsilon)]$$
$$+ \int_{\varepsilon}^{b} [f_x \phi_\varepsilon(t, \varepsilon) + f_r \phi_{t\varepsilon}(t, \varepsilon)] \, dt, \qquad (5.4)$$

where t, $\phi(t, \varepsilon)$, and $\phi_t(t, \varepsilon)$ are the arguments of f_x and f_r. Next, on invoking $\phi_{\varepsilon t} = \phi_{t\varepsilon}$ and requiring $f(\cdot)$ to be of class C^2, we integrate the second part of the integral in (5.4) by parts; namely,

$$\int_{\varepsilon}^{b} f_r \phi_{\varepsilon t}(t, \varepsilon) \, dt = f_r[t, \phi(t, \varepsilon), \phi_t(t, \varepsilon)] \phi_\varepsilon(t, \varepsilon) \Big|_{\varepsilon}^{b}$$
$$- \int_{\varepsilon}^{b} \left[\phi_\varepsilon(t, \varepsilon) \frac{d}{dt} f_r \right] dt$$

so that the entire integral in (5.4) becomes

$$\int_{\varepsilon}^{b} \left[f_x - \frac{d}{dt} f_r \right] \phi_\varepsilon(t, \varepsilon) \, dt + f_r[t, \phi(t, \varepsilon), \phi_t(t, \varepsilon)] \phi_\varepsilon(t, \varepsilon) \Big|_{\varepsilon}^{b}. \qquad (5.5)$$

Furthermore, from the definition of $\phi(t, \varepsilon)$, we have

$$\phi_\varepsilon(t, \varepsilon) = \frac{\dot{X}(\varepsilon) - \dot{x}^*(\varepsilon)}{b - \varepsilon}(b - t) + \frac{X(\varepsilon) - x^*(\varepsilon)}{(b - \varepsilon)^2}(b - t)$$

so that

$$\phi_\varepsilon(b, a) = 0, \qquad \phi_\varepsilon(a, a) = \dot{X}(a) - \dot{x}^*(a). \qquad (5.6)$$

Also,

$$\phi(t, a) = x^*(t) \quad \text{and} \quad \phi_t(t, a) = \dot{x}^*(t) \qquad \text{for } t \in [a, b]. \qquad (5.7)$$

Thus, letting $\varepsilon = a$ in (5.4) leads to (5.5) becoming

$$-f_r[a, x^*(a), \dot{x}^*(a)][\dot{X}(a) - \dot{x}^*(a)]$$

as a consequence of (5.6), (5.7), and Theorem 2.2, namely

$$f_x[t, x^*(t), \dot{x}^*(t)] - \frac{d}{dt} f_r[t, x^*(t), \dot{x}^*(t)] = 0,$$

so that

$$\Phi'(a) = f[a, X(a), \dot{X}(a)] - f[a, \phi(a, a), \phi_t(a, a)]$$
$$- [\dot{X}(a) - \dot{x}*(a)] f_r[a, x*(a), \dot{x}*(a)]. \quad (5.8)$$

However,

$$\phi(a, a) = x*(a), \qquad \phi_t(a, a) = \dot{x}*(a),$$
$$X(a) = x*(a), \qquad \dot{X}(a) = q,$$

so that the condition (5.3), together with (5.1) and (5.8), results in

$$\Phi'(a) = E[a, x*(a), \dot{x}*(a), q] \geq 0 \quad (5.9)$$

for all $q \in R^1$.

This proves the theorem for all $a \in [t_0, t_1)$ other than discontinuity points of $\dot{x}*(\cdot)$. By the assumed smoothness of $f(\cdot)$, $E[t, x*(t), \dot{x}*(t), q]$ is the value of a continuous function of t on subintervals between successive corners of $x*(\cdot)$. To establish the condition (5.2) for all $t \in [t_0, t_1]$, we let $t \to t_1$ from below, and $t \to t_c$ from below and from above if $\dot{x}*(\cdot)$ is discontinuous at $t_c \in (t_0, t_1)$. □

If $x*(\cdot)$ furnishes a strong local *maximum* (rather than minimum), the inequality sign in the condition (5.2) is reversed.

5.3. Example

Recall Example 3.3; namely,

$$f(t, x, r) = r^2$$

with the end conditions

$$(t_0, x_0) = (0, 0), \qquad (t_1, x_1) = (1, 1).$$

We found that the function with values

$$x*(t) = t$$

is extremal but does *not* furnish a global maximum of the integral $J(\cdot)$. Employing Theorem 5.1 (modified for a maximum), we show now that it does not even furnish

Chap. 5 • The Weierstrass Necessary Condition

a strong local maximum. Here

$$E[t, x^*(t), \dot{x}^*(t), q] = q^2 - 1 - 2(q-1)$$
$$= (q-1)^2,$$

which is *positive* for all $q \neq 1$.

5.4. The Legendre Necessary Condition

The Weierstrass condition embodied in Theorem 5.1 has an immediate corollary, the so-called Legendre[†] condition. This additional necessary condition is obtained from the proof of the Weierstrass condition by restricting the function $x(\cdot)$ to belong to a δ-neighborhood of order 1 of $x^*(\cdot)$, $x(\cdot) \in \mathcal{X} \cap N_1[\delta, x^*(\cdot)]$. The condition of the next theorem is necessary if $x^*(\cdot)$ renders a *weak* local minimum of $J(\cdot)$; hence it is also necessary for a strong local as well as for a global minimum.

Theorem 5.2. *If $x^*(\cdot) \in \mathcal{X}$ furnishes a weak local minimum of $J(\cdot)$ on $\mathcal{X}$, then*

$$f_{rr}[t, x^*(t), \dot{x}^*(t)] \geq 0 \qquad (5.10)$$

for all $t \in [t_0, t_1]$. If $\dot{x}^*(\cdot)$ is discontinuous at $t \in [t_0, t_1]$, this condition applies for $\dot{x}^*(t-0)$ and $\dot{x}^*(t+0)$.

As in Theorem 5.1, the inequality sign is reversed if $x^*(\cdot)$ furnishes a maximum.

Proof. Given $x^*(\cdot)$ and a $\delta > 0$, consider the function $x(\cdot)$ defined in the proof of Theorem 5.1, but now with the restriction $x(\cdot) \in \mathcal{X} \cap N_1[\delta, x^*(\cdot)]$. This latter condition implies for given $\varepsilon > 0$ that

$$\sup_{t \in [a, \varepsilon]} |X(t) - x^*(t)| + \sup_{t \in [a, \varepsilon]} |\dot{X}(t) - \dot{x}^*(t)| < \delta. \qquad (5.11)$$

Since $X(t) \to x^*(t)$ as $\varepsilon \to a$, the condition (5.11) implies that

$$|\dot{X}(a) - \dot{x}^*(a)| < \delta \qquad \text{for } \varepsilon \to a.$$

[†]A. M. Legendre, French mathematician, 1752–1833.

Thus, as in the proof of Theorem 5.1, we arrive at the condition (5.9) modified to

$$E[a, x^*(a), \dot{x}^*(a), q] \geq 0 \qquad (5.12)$$

for all q such that $|q - \dot{x}^*(a)| < \delta$. Again this condition holds for all $a \in [t_0, t_1)$, where $x^*(\cdot)$ is discontinuous.

Now we apply Taylor's formula with remainder (Ref. 5.1); namely,

$$f[a, x^*(a), q] = f[a, x^*(a), \dot{x}^*(a)] + [q - \dot{x}^*(a)] f_r[a, x^*(a), \dot{x}^*(a)]$$
$$+ \tfrac{1}{2}[q - \dot{x}^*(a)]^2 f_{rr}[a, x^*(a), \dot{x}^*(a) + \theta(q - \dot{x}^*(a))]$$

where $\theta \in (0, 1)$, so that

$$E[a, x^*(a), \dot{x}^*(a), q] = \tfrac{1}{2}[q - \dot{x}^*(a)]^2 f_{rr}[a, x^*(a), \dot{x}^*(a) + \theta(q - \dot{x}^*(a))]$$

and hence, by (5.12),

$$f_{rr}[a, x^*(a), \dot{x}^*(a) + \theta(q - \dot{x}^*(a))] \geq 0 \qquad (5.13)$$

for all q such that $0 < |q - \dot{x}^*(a)| < \delta$.

Since $f(\cdot)$ is assumed to be of class C^2 so that $f_{rr}(\cdot)$ is continuous in r, the condition (5.13) implies that

$$f_{rr}[a, x^*(a), \dot{x}^*(a)] \geq 0.$$

The condition at t_1 and at corners follows again by continuity. □

5.5. Example

Again recall Example 3.3, namely,

$$f(t, x, r) = r^2$$

with the endpoints

$$(t_0, x_0) = (0, 0), \qquad (t_1, x_1) = (1, 1).$$

We have already shown that the extremal $x^*(\cdot)$ such that

$$x^*(t) = t$$

Chap. 5 • The Weierstrass Necessary Condition

furnishes neither a global nor a strong local maximum. Since

$$f_{rr}(t,x,r)=2$$

for all $[t \ x \ r]^T \in [t_0,t_1] \times R^2$, and

$$f_{rr}[t,x^*(t),\dot{x}^*(t)] \leq 0$$

is necessary for a weak local maximum, $x^*(\cdot)$ does not even furnish a weak local maximum.

Exercises

5.1. Consider the integral $J(\cdot)$ with the integrand $f(t,x,r)=x^2(1-r)^2$. Deduce the excess function $E(t,x,r,q)$. Show that no extremal furnishes a strong or weak local maximum if $x_0 \neq 0$ or $x_1 \neq 0$.

5.2. Consider the integral $J(\cdot)$ with the integrand $f(t,x,r)=r^2-x^2$. Deduce the excess function $E(t,x,r,q)$. Show that no extremal furnishes a strong or weak local maximum regardless of end conditions.

5.3. Consider the integral $J(\cdot)$ with the integrand $f(t,x,r)=(r^2-k^2x^2)\exp(2t)$, $k=$ constant >1. Deduce the excess function $E(t,x,r,q)$. Show that no extremal furnishes a strong or weak local maximum regardless of end conditions.

5.4. Show that every extremal for the integrals in Exercises 5.1–5.3 satisfies the necessary conditions for a strong and weak local minimum, that is, the conditions of Weierstrass and Legendre, respectively.

5.5. Consider the integrals whose integrands are linear in r, that is, $f(t,x,r) = M(t,x)+N(t,x)r$. Show that the necessary conditions for a local extremum—strong or weak local minimum or maximum—namely, the conditions of Weierstrass and Legendre, respectively, are satisfied.

5.6. Consider the integral $J(\cdot)$ with the integrand $f(t,x,r)=x^2(1-r^2)$. Show that
  (a) the Weierstrass and Legendre necessary conditions for local maxima are satisfied by all extremals, and
  (b) no extremal furnishes a local minimum, strong or weak, if $x_0 \neq 0$ or $x_1 \neq 0$.

5.7. Consider the integral $J(\cdot)$ with the integrand $f(t,x,r)=r^3$, and the end points $(t_0,x_0)=(0,0),(t_1,x_1)=(1,1)$. Show that $x^*(\cdot)$ such that $x^*(t)=t$ is an extremal that satisfies the Legendre but not the Weierstrass condition for a local minimum.

5.8. Consider the integral $J(\cdot)$ with the integrand $f(t,x,r)=(1+r^2)^{1/2}/x$ and the end conditions such that $x_0>0, x_1>0$. Show that all extremals, meeting such end conditions, satisfy the necessary conditions of Weierstrass and of Legendre for a local minimum.

6

Jacobi's Necessary Condition

6.1. Introduction

Thus far we have considered three necessary conditions for a function $x^*(\cdot)$ to render a minimum of an integral $J(\cdot)$ on $\mathcal{X}$. These conditions are the Euler, Weierstrass, and Legendre conditions.

Since $\mathcal{X}$ is a class of piecewise smooth functions, extremals—that is, functions satisfying the Euler necessary condition and, of course, the prescribed end conditions—may have corners, that is, discontinuous derivatives. Now we turn to a necessary condition for a minimum (global, strong, or weak local) that must be met by *smooth* extremals, that is, by extremals without corners; this is *Jacobi's*[†] *necessary condition*.

6.2. The Accessory Minimum Problem

If $x^*(\ )$ furnishes a minimum (global or local) of integral $J(\cdot)$ on $\mathcal{X}$, then the twice-differentiable function $F(\cdot)$ in Section 2.6 possesses a minimum at $\varepsilon = 0$. Thus, in accordance with conditions (2.5), it is necessary that

$$F'(0) = 0 \quad \text{and} \quad F''(0) \geq 0. \tag{6.1}$$

The first of these conditions was utilized to deduce the Euler necessary condition, Theorem 2.1. Now we shall employ the second of conditions (6.1) to derive another necessary condition.

[†] C. G. J. Jacobi, German mathematician, 1804–1851.

On differentiating $F(\cdot)$, invoking $f_{xr} = f_{rx}$ which requires that $f(\cdot)$ be of class C^2, and letting $\varepsilon \to 0$, we obtain

$$F''(0) = \int_{t_0}^{t_1} \left[f_{xx}\eta^2(t) + 2f_{xr}\eta(t)\dot{\eta}(t) + f_{rr}\dot{\eta}^2(t) \right] dt \qquad (6.2)$$

where t, $x^*(t)$, and $\dot{x}^*(t)$ are the arguments of f_{xx}, f_{xr}, and f_{rr}. For a given minimizing function $x^*(\cdot)$, we denote the integrand on the right side of (6.2) by

$$2\omega[t, \eta(t), \dot{\eta}(t)] \triangleq f_{xx}\eta^2(t) + 2f_{xr}\eta(t)\dot{\eta}(t) + f_{rr}\dot{\eta}^2(t).$$

Since the integral in (6.2) depends on the choice of the function $\eta(\cdot)$, we denote it by

$$\mathcal{J}(\eta(\cdot)) \triangleq \int_{t_0}^{t_1} 2\omega[t, \eta(t), \dot{\eta}(t)] dt. \qquad (6.3)$$

Now let $\mathcal{K}$ denote the class of all piecewise smooth $\eta(\cdot): [t_0, t_1] \to R^1$ satisfying $\eta(t_0) = \eta(t_1) = 0$. Since the condition (6.1) must be met for all $\eta(\cdot) \in \mathcal{K}$, it follows that

$$\mathcal{J}(\eta(\cdot)) \geq 0 \qquad \forall \eta(\cdot) \in \mathcal{K}. \qquad (6.4)$$

Furthermore, in view of (6.2),

$$\mathcal{J}(\eta(\cdot)) = 0 \qquad \text{if } \eta(t) \equiv 0.$$

Thus, $\mathcal{J}(\eta(\cdot)) = 0$ is the minimum of $\mathcal{J}(\cdot)$ on $\mathcal{K}$. In other words, if $x^*(\cdot)$ furnishes a minimum of $J(\cdot)$ on $\mathcal{X}$, then the minimum of $\mathcal{J}(\cdot)$ on $\mathcal{K}$ is zero. We shall investigate the consequences of this *necessary* condition for $x^*(\cdot)$ to be a minimizing function of $J(\cdot)$.

The problem of minimizing $\mathcal{J}(\cdot)$ on $\mathcal{K}$ is termed the *accessory minimum problem*. Before treating this problem, we present a useful preliminary lemma.

Lemma 6.1. *If $f_{rr}(t, x, r) \neq 0$ for all $[t \; x \; r]^T \in (t_0, t_1) \times R^2$, then no function $x^*(\cdot)$ having a corner can minimize or maximize $J(\cdot)$ on $\mathcal{X}$.*

Chap. 6 • Jacobi's Necessary Condition

Proof. Suppose that $J(x^*(\cdot))$ is a global or local minimum or maximum of $J(\cdot)$ on $\mathcal{X}$ and that $[t_c, x^*(t_c)]$ is a corner—that is, $\dot{x}^*(\cdot)$ is discontinuous at $t_c \in (t_0, t_1)$. By Theorem 2.1, which applies because the extremizing function $x^*(\cdot)$ must satisfy the Euler necessary condition,

$$f_r[t_c, x^*(t_c), p] - f_r[t_c, x^*(t_c), q] = 0 \tag{6.5}$$

where

$$p \stackrel{\triangle}{=} \dot{x}^*(t_c - 0), \qquad q \stackrel{\triangle}{=} \dot{x}^*(t_c + 0),$$

and $p \neq q$.

By the mean value theorem, the left side of (6.5) can be expressed as

$$(p-q) f_{rr}[t_c, x^*(t_c), q + \theta(p-q)], \qquad \theta \in (0, 1),$$

so that (6.5), together with the hypothesis that $f_{rr}(t, x, r) \neq 0$ for all $[t \; x \; r]^T$, implies that $p = q$, which contradicts the supposition of a corner. $\square$

Now let us return to the accessory minimum problem and let us apply Euler's necessary condition to this problem. If the hypotheses of Theorem 2.3 are met, then a function $\eta^*(\cdot)$ that minimizes $\mathcal{J}(\cdot)$ on $\mathcal{H}$ must be a solution of the corresponding Euler–Lagrange equation

$$\omega_\eta = \omega_{\dot{\eta}t} + \omega_{\dot{\eta}\eta}\dot{\eta}(t) + \omega_{\dot{\eta}\dot{\eta}}\ddot{\eta}(t) \tag{6.6}$$

where t, $\eta(t)$, and $\dot{\eta}(t)$ are the arguments of ω_η, $\omega_{\dot{\eta}t}$, $\omega_{\dot{\eta}\eta}$, and $\omega_{\dot{\eta}\dot{\eta}}$. Equation (6.6) is called *Jacobi's equation*. On expanding ω_η, $\omega_{\dot{\eta}t}$, $\omega_{\dot{\eta}\eta}$, and $\omega_{\dot{\eta}\dot{\eta}}$, it is readily seen that Jacobi's equation is a linear differential equation of second order with variable coefficients; we shall discuss its solutions in the next section. To ensure that Theorem 2.3 is valid, it suffices that the solutions of (6.6) are smooth—that is, have no corners—and that

$$\omega_{\dot{\eta}\dot{\eta}}[t, \eta^*(t), \dot{\eta}^*(t)] \neq 0 \qquad \forall t \in [t_0, t_1]. \tag{6.7}$$

To guarantee these conditions, we impose two conditions on $x^*(\cdot)$. For the remaining discussion in this chapter, we suppose that $x^*(\cdot)$ is smooth so that $\dot{x}^*(\cdot)$ is continuous on $[t_0, t_1]$, and

$$f_{rr}[t, x^*(t), \dot{x}^*(t)] > 0 \qquad \forall t \in [t_0, t_1]. \tag{6.8}$$

Condition (6.8) is termed the *strengthened Legendre condition*.

By differentiation of $\omega(\cdot)$, it follows that

$$\omega_{\dot{\eta}\dot{\eta}}[t,\eta(t),\dot{\eta}(t)] = f_{rr}[t, x^*(t), \dot{x}^*(t)] \tag{6.9}$$

for *any* $\eta(\cdot) \in \mathcal{H}$. Thus, in view of (6.8), the condition (6.7) is satisfied. Furthermore, since (6.9) holds for any $\eta(\cdot) \in \mathcal{H}$, it follows from (6.8) that

$$\omega_{\dot{\eta}\dot{\eta}}(t,\eta,\dot{\eta}) \neq 0 \quad \forall [t\ \eta\ \dot{\eta}]^T \in (t_0, t_1) \times R^2.$$

Hence, Lemma 6.1 employed for the accessory minimum problem guarantees that no function $\eta^*(\cdot)$ minimizing $\mathcal{J}(\cdot)$ on $\mathcal{H}$ can have a corner.

Before stating another necessary condition for $x^*(\cdot)$ to furnish a minimum of $J(\cdot)$, we introduce some definitions. If $\mu(\cdot): [t_0, t_2] \to R^1$ is a solution of Jacobi's equation (6.6) with $\mu(t_0) = \mu(t_2) = 0$ and $\mu(t) \neq 0$ for $t \in (t_0, t_2)$, then t_2 is termed a *conjugate value to t_0* and $[t_2, x^*(t_2)]$ is called a *conjugate point to the initial point* $[t_0, x^*(t_0)]$.

With these definitions we are ready to state *Jacobi's necessary condition*.

Theorem 6.1. *If $x^*(\cdot)$ is a smooth function that furnishes a minimum (or maximum) of $J(\cdot)$ on $\mathcal{X}$ and if the condition (6.8) holds (with the inequality reversed for a maximum), then there is no value t_2 conjugate to t_0 with $t_2 < t_1$.*

Proof. Suppose there is a value $t_2 < t_1$ conjugate to t_0; that is, there is a solution $\mu(\cdot)$ of (6.6) with $\mu(t_0) = \mu(t_2) = 0$ and $\mu(t) \neq 0$ on (t_0, t_2).

Now define a function $\eta^*(\cdot) \in \mathcal{H}$ by

$$\eta^*(t) = \begin{cases} \mu(t) & \text{for } t \in [t_0, t_2], \\ 0 & \text{for } t \in (t_2, t_1]. \end{cases} \tag{6.10}$$

Note that the function $\eta^*(\cdot)$ defined by (6.10) satisfies Jacobi's equation—that is, the Euler–Lagrange equation for the accessory minimum problem—in the form

$$\omega_\eta[t, \eta^*(t), \dot{\eta}^*(t)] = \frac{d}{dt} \omega_{\dot{\eta}}[t, \eta^*(t), \dot{\eta}^*(t)] \tag{6.11}$$

on $[t_0, t_2)$ and $(t_2, t_1]$, respectively.

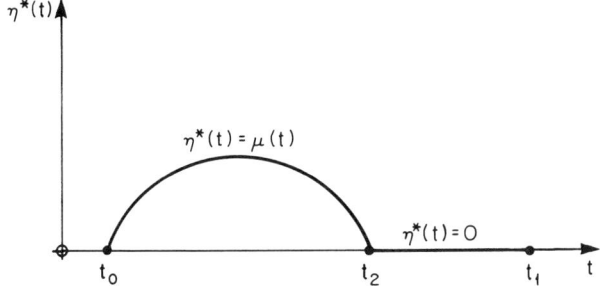

Figure 6.1. The function $\eta^*(\cdot)$.

As can be readily verified,

$$2\omega(t, \eta, \dot{\eta}) = \eta \omega_\eta(t, \eta, \dot{\eta}) + \dot{\eta} \omega_{\dot{\eta}}(t, \eta, \dot{\eta}). \tag{6.12}$$

On letting $\eta = \eta^*(t)$ and substituting the right side of (6.11) for ω_η in (6.12), we have

$$2\omega[t, \eta^*(t), \dot{\eta}^*(t)] = \eta^*(t) \frac{d}{dt} \omega_{\dot{\eta}}[t, \eta^*(t), \dot{\eta}^*(t)] + \dot{\eta}^*(t) \omega_{\dot{\eta}}[t, \eta^*(t), \dot{\eta}^*(t)]$$

$$= \frac{d}{dt} \eta^*(t) \omega_{\dot{\eta}}[t, \eta^*(t), \dot{\eta}^*(t)] \tag{6.13}$$

on $[t_0, t_2]$ and $[t_2, t_1]$, where $\dot{\eta}^*(t_2 - 0)$ and $\dot{\eta}^*(t_2 + 0)$ are defined but possibly different from each other. However, in view of the hypothesized smoothness of $x^*(\cdot)$ and the condition (6.8), it follows from Theorem 2.3 that $x^*(\cdot)$ is of class C^2; hence the right side of (6.13) is continuous on $[t_0, t_2]$ and $[t_2, t_1]$. Now we may write (6.3) as

$$\mathcal{J}(\eta^*(\cdot)) = \int_{t_0}^{t_2} d\{\eta^*(t) \omega_{\dot{\eta}}[t, \eta^*(t), \dot{\eta}^*(t)]\} + \int_{t_2}^{t_1} 2\omega[t, \eta^*(t), \dot{\eta}^*(t)] \, dt.$$

The first integral vanishes because $\eta^*(t_0) = \eta^*(t_2) = 0$. The second integral is zero because $\eta^*(t) \equiv 0$ on $[t_2, t_1]$ implies that $\omega[t, \eta^*(t), \dot{\eta}^*(t)] \equiv 0$. But, as noted earlier, the minimum of $\mathcal{J}(\cdot)$ on $\mathcal{H}$ is zero; hence $\eta^*(\cdot)$ minimizes $\mathcal{J}(\cdot)$ on $\mathcal{H}$.

We also noted earlier that a function which minimizes $\mathcal{J}(\cdot)$ on $\mathcal{H}$ cannot have a corner in view of condition (6.8). Thus it follows that

$$\dot{\eta}^*(t_2 - 0) = \dot{\eta}^*(t_2 + 0) = 0$$

so that, as a consequence of (6.10),

$$\dot{\mu}(t_2) = 0$$

as well.

Now, Jacobi's equation (6.6) is a homogeneous linear differential equation with continuous coefficients and with a nonvanishing coefficient of the highest-order derivative, $\ddot{\eta}(t)$. Thus, given the pair $[\eta(t_2), \dot{\eta}(t_2)] = [\mu(t_2), \dot{\mu}(t_2)]$, the solution $\mu(\cdot)$ is unique; for instance, see Ref. 6.1. In particular, if $\mu(t_2) = \dot{\mu}(t_2) = 0$, then $\mu(t) \equiv 0$ on $[t_0, t_2]$. But this is contrary to the hypothesis that $\mu(t) \neq 0$ for $t \in (t_0, t_2)$. Thus, t_2 cannot be conjugate to t_0 as supposed at the outset. □

6.3. The Integration of Jacobi's Equation

To verify whether or not there exists a value $t_2 < t_1$ conjugate to t_0, we need the general solution of Jacobi's equation (6.6). The general solution $\eta(\cdot): [t_0, t_1] \to R^1$ of a homogeneous linear differential equation such as (6.6) is given by

$$\eta(t) = c_1 \eta_1(t) + c_2 \eta_2(t), \qquad t \in [t_0, t_1] \tag{6.14}$$

where c_1 and c_2 are real numbers, and $\eta_1(\cdot)$ and $\eta_2(\cdot)$ are any two *linearly independent* solutions of (6.6); again, see Ref. 6.1.

A necessary and sufficient condition for the linear independence of $\eta_1(\cdot)$ and $\eta_2(\cdot)$ is the nonvanishing of the so-called Wronskian determinant on $[t_0, t_1]$; namely,

$$\begin{vmatrix} \eta_1(t) & \eta_2(t) \\ \dot{\eta}_1(t) & \dot{\eta}_2(t) \end{vmatrix} \neq 0 \qquad \forall t \in [t_0, t_1]. \tag{6.15}$$

If we have any two particular solutions, $\eta_1(\cdot)$ and $\eta_2(\cdot)$, which are linearly independent, and hence have the general solution $\eta(\cdot)$, we impose

$$\eta(t_0) = c_1 \eta_1(t_0) + c_2 \eta_2(t_0) = 0 \tag{6.16}$$

since we wish to investigate the existence of a value t_2 conjugate to t_0. Thus we have

$$\frac{c_1}{c_2} = -\frac{\eta_2(t_0)}{\eta_1(t_0)}. \tag{6.17}$$

Chap. 6 • Jacobi's Necessary Condition

Next we prove another useful lemma.

Lemma 6.2. *The value t_2 conjugate to t_0 is independent of the particular choice of a solution of Jacobi's equation.*

Proof. We shall prove the lemma by showing that any two solutions vanishing at t_0 differ from each other by a nonzero constant factor.

Let $\eta_a(\cdot)$ and $\eta_b(\cdot)$ be two distinct solutions of (6.6), such that $\eta_a(t_0) = \eta_b(t_0) = 0$ but neither vanishes identically on $[t_0, t_1]$. Hence $\dot\eta_a(t_0) \neq 0$ and $\dot\eta_b(t_0) \neq 0$, so that there is a nonzero constant k such that

$$\dot\eta_a(t_0) - k\dot\eta_b(t_0) = 0.$$

Since $\eta_a(\cdot)$ and $\eta_b(\cdot)$ are solutions, so is the sum $\eta_a(\cdot) - k\eta_b(\cdot)$ which, together with its derivative, vanishes at t_0; consequently, this solution vanishes identically on $[t_0, t_1]$. It follows that

$$\eta_a(t) = k\eta_b(t) \qquad \forall t \in [t_0, t_1], \quad k \neq 0.$$

Hence we conclude that

$$\eta_a(t) \neq 0 \Leftrightarrow \eta_b(t) \neq 0$$

and

$$\eta_a(t_2) = 0 \Leftrightarrow \eta_b(t_2) = 0,$$

and so a conjugate value t_2 of one is also a conjugate value of the other. □

Now, returning to (6.14)–(6.17), we see that Lemma 6.2 permits us to utilize *any* solution $\eta(\cdot)$ with $\eta(t_0) = 0$ to determine the existence of *the* value t_2 conjugate to t_0; that is, we may pick any one of the one-parameter family of solutions (6.14) with (6.17). In other words, we may impose another condition on c_1 and c_2; for instance we may let $c_2 = -\eta_1(t_0)$, so that

$$\eta(t) = \eta_2(t_0)\eta_1(t) - \eta_1(t_0)\eta_2(t). \tag{6.18}$$

This particular solution can be employed to check for the existence of a conjugate value t_2. To construct $\eta(\cdot)$ given by (6.18), we require two linearly independent solutions, $\eta_1(\cdot)$ and $\eta_2(\cdot)$. The following lemma is useful for that purpose.

Lemma 6.3. *If* $g(\cdot):[t_0,t_1]\times R^2 \to R^1$ *is the general solution of the Euler–Lagrange equation (2.23), so that* $x(t)=g(t,\alpha,\beta)$ *for* $[\alpha\ \beta]^T \in R^2$, *and* α^*, β^* *are the values of* α, β *such that*

$$x_0 = g(t_0, \alpha^*, \beta^*), \qquad x_1 = g(t_1, \alpha^*, \beta^*)$$

then $\eta_1(\cdot)$ *and* $\eta_2(\cdot)$ *with the values*

$$\eta_1(t) = g_\alpha(t, \alpha^*, \beta^*), \qquad \eta_2(t) = g_\beta(t, \alpha^*, \beta^*)$$

are two linearly independent solutions of Jacobi's equation (6.6).

Proof. First of all consider Jacobi's equation (6.6). Let

$$P(t) \triangleq f_{xx}[t, x^*(t), \dot{x}^*(t)],$$

$$Q(t) \triangleq f_{xr}[t, x^*(t), \dot{x}^*(t)],$$

$$R(t) \triangleq f_{rr}[t, x^*(t), \dot{x}^*(t)]$$

so that the integrand in (6.2) becomes

$$2\omega[t, \eta(t), \dot{\eta}(t)] = P(t)\eta^2(t) + 2Q(t)\eta(t)\dot{\eta}(t) + R(t)\dot{\eta}^2(t).$$

Then, writing Jacobi's equation in the form of (6.11), it becomes

$$[P(t) - \dot{Q}(t)]\eta(t) - \frac{d}{dt}[R(t)\dot{\eta}(t)] = 0. \qquad (6.19)$$

Thus, (6.19) is one way of writing Jacobi's equation (6.6).

Under the hypotheses which validate Euler's necessary condition, Theorem 2.3, in terms of Euler–Lagrange equation (2.23) on $[t_0, t_1]$, that equation possesses a two-parameter family of solutions with values

$$x(t) = g(t, \alpha, \beta), \qquad [t\ \alpha\ \beta]^T \in [t_0, t_1] \times R^2$$

one of which is

$$x^*(t) = g(t, \alpha^*, \beta^*)$$

and such that the partial derivatives g_α, g_β, g_t, $g_{\alpha t}$, and $g_{\beta t}$ exist and are

Chap. 6 • Jacobi's Necessary Condition

continuous for $t \in [t_0, t_1]$ and α, β sufficiently near α^*, β^*; for instance, see Ref. 6.1.

On substituting the general solution in the form (2.22) of the Euler-Lagrange equation, one has

$$f_x[t, g(t, \alpha, \beta), g_t(t, \alpha, \beta)] - \frac{d}{dt} f_{\dot{x}}[t, g(t, \alpha, \beta), g_t(t, \alpha, \beta)] = 0$$

for all $[t \ \alpha \ \beta]^T \in [t_0, t_1] \times R^2$. Since this relation holds identically, the derivatives of the left side with respect to t, α, and β also vanish. Upon differentiation with respect to α and β, and setting $g_{\alpha t} = g_{t\alpha}$ and $g_{\beta t} = g_{t\beta}$, one obtains

$$\left(f_{xx} - \frac{d}{dt}f_{\dot{x}x}\right)g_\alpha - \frac{d}{dt}(f_{\dot{x}\dot{x}}g_{\alpha t}) = 0,$$

$$\left(f_{xx} - \frac{d}{dt}f_{\dot{x}x}\right)g_\beta - \frac{d}{dt}(f_{\dot{x}\dot{x}}g_{\beta t}) = 0, \qquad (6.20)$$

where the arguments of f_{xx}, $f_{\dot{x}x}$, and $f_{\dot{x}\dot{x}}$ are t, g, g_t, and those of g, g_t, g_α, g_β, $g_{\alpha t}$, and $g_{\beta t}$ are t, α, β.

Recalling the definitions of $P(t)$, $Q(t)$, and $R(t)$, and setting $\alpha = \alpha^*$, $\beta = \beta^*$ in (6.20), we arrive at

$$[P(t) - \dot{Q}(t)]g_\alpha - \frac{d}{dt}[R(t)g_{\alpha t}] = 0,$$

$$[P(t) - \dot{Q}(t)]g_\beta - \frac{d}{dt}[R(t)g_{\beta t}] = 0. \qquad (6.21)$$

Each of these is Jacobi's equation in the form (6.19); that is, $\eta_1(\cdot)$ and $\eta_2(\cdot)$ with the values

$$\eta_1(t) = g_\alpha(t, \alpha^*, \beta^*), \qquad \eta_2(t) = g_\beta(t, \alpha^*, \beta^*)$$

are solutions of Jacobi's equation. It remains to show that they are linearly independent.

If, as we suppose, $g(\cdot)$ is the general solution of the Euler-Lagrange equation, then the so-called Jacobian determinant must be nonzero (for instance, see Ref. 6.1); namely,

$$\begin{vmatrix} g_\alpha & g_{t\alpha} \\ g_\beta & g_{t\beta} \end{vmatrix} \neq 0 \qquad (6.22)$$

for all $t \in [t_0, t_1]$ and all α, β sufficiently near α^*, β^*. Setting $g_{t\alpha} = g_{\alpha t}$ and $g_{t\beta} = g_{\beta t}$ and $\alpha = \alpha^*$, $\beta = \beta^*$ reduces (6.22) to (6.15); thus, $\eta_1(\cdot)$ and $\eta_2(\cdot)$ are linearly independent. $\square$

6.4. Example

Recall once again Example 3.3; namely, the integrand is

$$f(t, x, r) = r^2$$

and the end points are

$$(t_0, x_0) = (0, 0), \qquad (t_1, x_1) = (1, 1).$$

As we showed earlier, $x^*(\cdot)$ given by $x^*(t) = t$ is an extremal. Does it satisfy Jacobi's necessary condition?

To check for a conjugate value $t_2 < 1$, we form

$$\eta(t) = \eta_2(t_0)\eta_1(t) - \eta_1(t_0)\eta_2(t)$$

with

$$\eta_1(t) = g_\alpha(t, \alpha^*, \beta^*), \qquad \eta_2(t) = g_\beta(t, \alpha^*, \beta^*),$$

where

$$x(t) = g(t, \alpha, \beta) = \alpha + \beta t$$

and

$$x^*(t) = t \qquad (\alpha^* = 0, \beta^* = 1).$$

Consequently we obtain

$$g_\alpha(t, \alpha, \beta) = 1, \qquad g_\beta(t, \alpha, \beta) = t$$

for all α, β. Hence we arrive at

$$\eta(t) = (0)(1) - (1)(t) = -t,$$

which is nonzero for $t \neq t_0 = 0$. Thus there is no value t_2 conjugate to t_0 and Jacobi's *necessary* condition is met.

Exercises

6.1. Discuss the validity of Jacobi's necessary condition for integrals with integrands not depending on r, that is, with $f_r \equiv 0$.

6.2. Discuss the validity of Jacobi's necessary condition for the integral $J(\cdot)$ with the integrand $f(t, x, r) = r^2$ and end points such that $x_0 = x_1$.

6.3. Consider $J(\cdot)$ with $f(t, x, r) = r^2 + r$, and the end points $(t_0, x_0) = (0, 0)$ and $(t_1, x_1) = (1, 1)$. Show that the extremal corresponding to minimizing $J(\cdot)$ satisfies the necessary conditions of Weierstrass, Legendre, and Jacobi.

6.4. Consider the integral in Exercise 5.3 where $f(t, x, r) = (r^2 - k^2 x^2)\exp(2t)$, $k = \text{constant} > 1$. Deduce the general solution of the corresponding Euler–Lagrange equation. Show that Jacobi's necessary condition is not satisfied for $t_0 = 0$ and $t_1 > \pi(k^2 - 1)^{-1/2}$.

6.5. Consider integrals $J(\cdot)$ with integrands which do not depend on x, that is, $f_x \equiv 0$. Show that the corresponding extremals possess no points conjugate to the initial point.

6.6. Consider the integral $J(\cdot)$ with the integrand $f(t, x, r) = t^2 + x^2 + r^2$, and the end points such that $t_0 = 0$, $x_0 = x_1 = 0$. Discuss Jacobi's necessary condition for all values of t_1.

6.7. Consider the problem of minimizing the surface of revolution generated by a curve that is revolved about a line. Show that the corresponding integral has the integrand $f(t, x, r) = x(1 + r^2)^{1/2}$. Discuss Jacobi's necessary condition for all possible end points of the curve.

7

Corner Conditions

7.1. Necessary Conditions

Thus far we have derived four sets of necessary conditions which must be met by a function $x^*(\cdot)$ that furnishes a minimum of the integral $J(\cdot)$ over the class of admissible functions $\mathcal{X}$; they are the conditions of Euler, Weierstrass, Legendre, and Jacobi. Now we consider conditions that must be met at a corner of $x^*(\cdot)$, that is, at a point $[t_c, x^*(t_c)]$, $t_c \in (t_0, t_1)$, where $\dot{x}^*(\cdot)$ is discontinuous. These conditions are often referred to as the Erdmann–Weierstrass *corner conditions*.[†] The first of them is embodied in the following theorem.

Theorem 7.1. *If $x^*(\cdot)$ furnishes a weak local minimum (or maximum) of $J(\cdot)$ on $\mathcal{X}$ and $[t_c, x^*(t_c)]$ is a corner, then*

$$f_i[t_c, x^*(t_c), \dot{x}^*(t_c - 0)] = f_{\dot{r}}[t_c, x^*(t_c), \dot{x}^*(t_c + 0)]. \tag{7.1}$$

Proof. The condition (7.1) follows at once from Theorem 2.1. The integral in (2.19) is a continuous function of t on $[t_0, t_1]$. The condition (7.1) is obtained by letting $t \to t_c$ from below and from above, respectively. □

Another necessary condition is a consequence of the Weierstrass condition and hence applies to strong local minima (or maxima).

[†]Erdmann gave an independent proof in 1877.

Theorem 7.2. *If $x^*(\cdot)$ furnishes a strong local minimum (or maximum) of $J(\cdot)$ on $\mathcal{X}$ and $[t_c, x^*(t_c)]$ is a corner, then*

$$f[t_c, x^*(t_c), \dot{x}^*(t_c-0)] - \dot{x}^*(t_c-0)f_r[t_c, x^*(t_c), \dot{x}^*(t_c-0)]$$
$$= f[t_c, x^*(t_c), \dot{x}^*(t_c+0)] - \dot{x}^*(t_c+0)f_r[t_c, x^*(t_c), \dot{x}^*(t_c+0)].$$
(7.2)

Proof. Apply the condition (5.2) in Theorem 5.1 with $\dot{x}^*(t) = \dot{x}^*(t_c - 0)$ and $q = \dot{x}^*(t_c + 0)$, as well as with $\dot{x}^*(t) = \dot{x}^*(t_c + 0)$ and $q = \dot{x}^*(t_c - 0)$, and $t = t_c$ and $x^*(t) = x^*(t_c)$ in both cases. The two resulting inequalities, together with the condition (7.1), imply (7.2). □

7.2. Example

Consider the integral $J(\cdot)$ with the integrand $f(t, x, r) = (r^2 - 1)^2$, so that

$$f_r = 4r(r^2 - 1), \qquad f_{rr} = 12r^2 - 4.$$

Thus the hypothesis of Lemma 6.1 is not met; that is, one cannot assert that $f_{rr} \neq 0$ for all $[t \; x \; r]^T$. Consequently one cannot rule out the occurrence of corners for a globally or locally minimizing function $x^*(\cdot)$.

By Euler's necessary condition, Theorem 2.1, if $x^*(\cdot)$ furnishes a minimum of $J(\cdot)$, then

$$\dot{x}^*(t)[\dot{x}^{*2}(t) - 1] = \text{constant} \qquad (7.3)$$

on $[t_0, t_1]$. Thus we can say that $\dot{x}^*(t) = \text{constant}$ between corners, but as yet we cannot assert more. If there is a corner at $t_c \in (t_0, t_1)$, then the corner conditions (7.1) and (7.2) require

$$p(p^2 - 1) = q(q^2 - 1),$$
$$(p^2 - 1)(3p^2 + 1) = (q^2 - 1)(3q^2 + 1), \qquad (7.4)$$

where

$$p \overset{\triangle}{=} \dot{x}^*(t_c - 0), \qquad q \overset{\triangle}{=} \dot{x}^*(t_c + 0).$$

The system (7.4) has the trivial solutions $p = q$; that is, there is no corner at t_c. It can be shown that the only nontrivial solutions are

$$p = 1, \; q = -1 \quad \text{and} \quad p = -1, \; q = 1. \qquad (7.5)$$

Chap. 7 • Corner Conditions

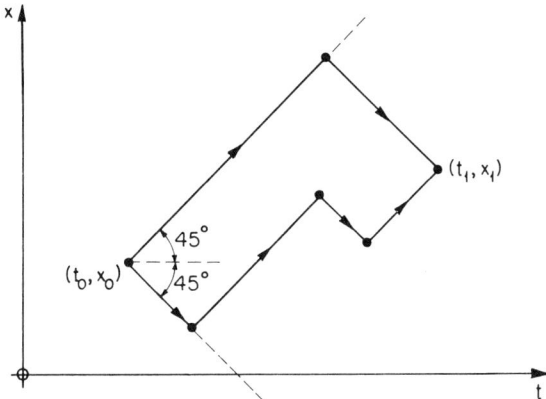

Figure 7.1. Extremals with corners.

Hence, if there is one or more than one corner, the function $\dot{x}^*(\cdot)$ is piecewise linear and successive segments have slope 1 or -1. If $x^*(\cdot)$ has no corner, it must have $\dot{x}^*(t) = $ constant on $[t_0, t_1]$.

Now it is easily seen by inspection of

$$J(x(\cdot)) = \int_{t_0}^{t_1} \left[\dot{x}^2(t) - 1 \right]^2 dt \tag{7.6}$$

that the infimum (greatest lower bound) of $J(\cdot)$ is zero. Thus, if the end points (t_0, x_0) and (t_1, x_1) can be joined by line segments of alternating slope 1 and -1,

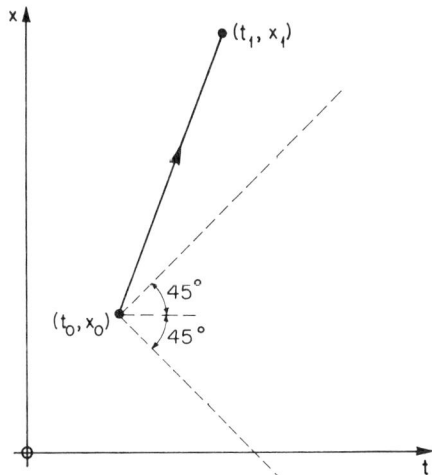

Figure 7.2. A smooth extremal.

then $x^*(\cdot)$ renders a global minimum of $J(\cdot)$ on $\mathfrak{X}$. However, it is readily seen that not all end points can be so joined. In fact, given (t_0, x_0), the end point (t_1, x_1) must lie in the conic sector contained within the lines of slopes 1 and -1 emanating from (t_0, x_0); see Figure 7.1. Endpoints so situated relative to each other can be joined by an infinite number of such minimizing curves with corners; two of them are shown in Figure 7.1.

If (t_1, x_1) lies outside the conic sector described above, and hence cannot be joined to (t_0, x_0) by a globally minimizing curve with corners, the only possible candidate for a minimizing function is an extremal without corners, that is, a line; see Figure 7.2. It is left as an exercise for the reader to show that such an extremal, $x^*(t) = \alpha + \beta t$, with slope β such that $\beta^2 > 1$, does satisfy the remaining *necessary* conditions of Weierstrass, Legendre, and Jacobi. Nonetheless, without further proof one cannot assert that such an extremal furnishes a minimum (of any kind) for $J(\cdot)$.

Exercises

7.1. Consider the integral $J(\cdot)$ with the integrand $f(t, x, r) = r^2(1+r)^2$, and the end points $(t_0, x_0) = (0, 0)$ and $(t_1, x_1) = (1, -\frac{1}{2})$. Using the corner conditions, deduce a globally minimizing function with corners.

7.2. Consider the integral $J(\cdot)$ with the integrand $f(t, x, r) = x^2(1-r)^2$, and the end conditions $x(t_0) = x_0$ and $x(t_1) = x_1$. Employing the corner conditions, discuss the possibility of extremals with corners.

7.3. Consider the integral $J(\cdot)$ with the integrand $f(t, x, r) = r^2 + xr + x^2$. Show that extremals must be smooth as a consequence of the corner conditions.

7.4. Consider the integral $J(\cdot)$ with the integrand $f(t, x, r) = (r-1)^2(r+1)^2$, and the end points $(t_0, x_0) = (0, 0)$ and $(t_1, x_1) = (2, 1)$. Deduce an extremal with a corner.

7.5. Consider the integral $J(\cdot)$ with the integrand $f(t, x, r) = r^2(1-r)^2$, and the end points $(t_0, x_0) = (0, 0)$ and $(t_1, x_1) = (2, 1)$. Deduce a globally minimizing function.

7.6. Can there be extremals with corners for the integral $J(\cdot)$ with $f(t, x, r) = r^2 - x^2$? Employ the corner conditions to furnish the answer.

7.7. Consider the integral $J(\cdot)$ with the integrand $f(t, x, r) = r^4 - 6r^2$. Discuss the existence of extremals with corners.

8

Concluding Remarks

In the aforegoing sections, Part I of this book, we consider the simplest problem of the calculus of variations, so called because it is indeed the simplest of a class of functional extremization problems. We restrict the discussion to functionals of integral type whose integrands depend on a scalar-valued variable $x(t)$ and its derivative $\dot{x}(t)$. Furthermore, we consider fixed end points; that is, we prescribe the initial and terminal values of t and $x(t)$. A number of generalizations are of interest and are introduced below; while not treated in Part I, many are in fact subsumed in the class of problems discussed in Part II.

One way of enlarging the class of problems consists of relaxing the end conditions. Instead of prescribing the initial and terminal values of t and $x(t)$, one can simply require that $[t_0, x(t_0)]$ and $[t_1, x(t_1)]$ lie on prescribed curves; see Figure 8.1; that is, given the functions $\theta^0(\cdot): R^2 \to R^1$ and $\theta^1(\cdot): R^2 \to R^1$, we require that the class of admissible functions $\mathcal{X}$ consist of all piecewise smooth functions $x(\cdot): [t_0, t_1] \to R^1$ such that

$$\theta^0[t_0, x(t_0)] = 0 \quad \text{and} \quad \theta^1[t_1, x(t_1)] = 0. \tag{8.1}$$

Problems of this type are subsumed in the class of problems treated in Part II.

In Section 3 we introduce the problem of extremizing the integral $J(\cdot)$ on a set of piecewise smooth *vector-valued* functions $x(\cdot): [t_0, t_1] \to R^n$ with fixed end points. Relaxing the end conditions as discussed above is obvious; in (8.1) the functions $\theta^0(\cdot)$ and $\theta^1(\cdot)$ are then $\theta^0(\cdot): R^{n+1} \to R^p$ and $\theta^1(\cdot): R^{n+1} \to R^q$, p and $q \leq n+1$. However, when $x(\cdot)$ is vector-valued, other generalizations are of interest. Thus one may require that admissible functions satisfy *constraints*. For instance, one may consider $\mathcal{X}$ to be the

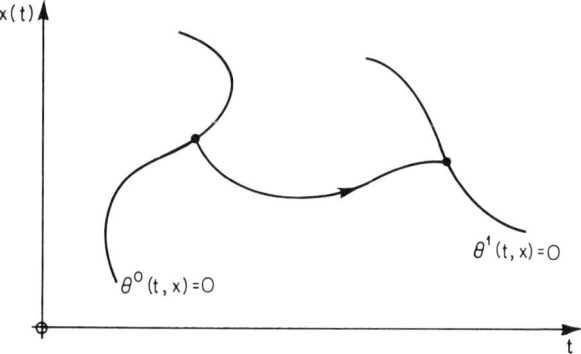

Figure 8.1. Variable end points.

class of all piecewise smooth functions $x(\cdot): R^1 \to R^n$ satisfying a *finite constraint*

$$\phi[t, x(t)] = 0 \tag{8.2}$$

where $\phi(\cdot): R^{n+1} \to R^k, k \leq n$, is prescribed, as well as certain end conditions, such as those of type (8.1); or one may impose a *differential constraint*

$$\psi[t, x(t), \dot{x}(t)] = 0 \tag{8.3}$$

where $\psi(\cdot): R^{2n+1} \to R^l, l \leq n$, is given. Of course, one may impose both (8.2) and (8.3).

Certain classes of problems with differential as well as with finite constraints, as well as with other types of constraints, are discussed in Part II. However, the point of view adopted in Part II is *geometric* in nature and hence quite different from that in Part I, which is *variational*. Readers interested in a variational approach to more general classes of extremization problems, such as the ones mentioned above, may wish to consult such standard texts as Ref. 8.1.

Another point worth mentioning concerns the scope of the conditions derived in Part I; they are *necessary* conditions for $J(\cdot)$ to take on an extremum. Conditions which are *sufficient* to assure an extremum are not discussed in Part I. This is done in Part II.

Before leaving Part I, devoted to the simplest problem, a word may be in order about restricting $\mathcal{X}$ to the class of piecewise smooth functions. Consider the integral

$$J(x(\cdot)) = \int_0^1 \left\{ x^2(t) + \left[\dot{x}^2(t) - 1 \right]^2 \right\} dt.$$

Chap. 8 • Concluding Remarks

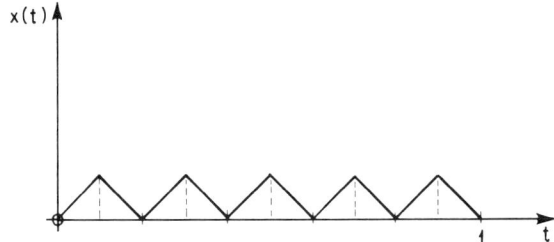

Figure 8.2. A piecewise linear function.

Since the lower bound of the integrand is zero, the infimum of $J(\cdot)$ cannot be less than zero. Can this zero value be achieved by a piecewise smooth function? To achieve $J(x(\cdot))=0$, one must have $x(t)=0$ on $[0,1]$ to assure the vanishing of the first term in the integrand. But then $\dot{x}(t)\equiv 0$ as well, and so the integral $J(x(\cdot))=1$, not $J(x(\cdot))=0$. Suppose we impose the end conditions $x(0)=x(1)=0$, and consider $[0,1]$ divided into n equal subintervals each of length $1/n$. Let $x(\cdot)$ be the piecewise linear function with the value

$$x(t) = t - \frac{k}{n} \qquad \text{for} \quad t \in \left[\frac{k}{n}, \frac{k}{n} + \frac{1}{2n}\right],$$

$$x(t) = -t + \frac{k+1}{n} \qquad \text{for} \quad t \in \left[\frac{k}{n} + \frac{1}{2n}, \frac{k+1}{n}\right]$$

for $k=0,1,\ldots,n-1$. This saw-tooth function is shown in Figure 8.2. With this $x(\cdot)\in\mathcal{X}$ the value of the integral is

$$J(x(\cdot)) = 1/12n^2.$$

Thus, by increasing the number of intervals, one can decrease the value of the integral and approach the infimum as closely as desired. However, as long as $x(\cdot)$ is restricted to be piecewise smooth, the infimum cannot be attained. It is for reasons such as this that a theory allowing for a larger class of admissible functions—namely, Lebesgue measurable ones—has been devised. Such a theory is beyond the scope of this book, which is devoted to what is sometimes dubbed the "naive" theory. For an enlightening discussion of this topic, the reader is referred to the introductory article in Ref. 8.2.

Part II

Optimal Control

9

Introduction

Recall here the problem of range maximization for a rocket plane in horizontal flight, posed in Chapter 1. Three variables, distance x, speed v, and mass m are used to describe the salient features of the physical system; these variables constitute the so-called *state* of the system. The evolution of the state with time t is determined by a set of ordinary differential equations (1.3). These equations involve another variable, namely, the thrust $c\beta$ or, since c is a given constant, merely the mass flow rate β. Thus, if $\beta(t)$ is specified on some time interval $[t_0, t_1]$, and if $\beta(\cdot)$ is a sufficiently well-behaved function, say piecewise continuous, then, for given initial values of the state variables, equations (1.3) possess a unique solution on some time interval containing t_0. The variable β is termed a *control* variable because its choice determines or controls the solutions of the *state equations* (1.3). Thus there are three kinds of variables: The independent variable, here time t; the control variable(s), here mass flow rate $\beta(\cdot)$; and the state variables, here distance x, speed v, and mass m. Loosely speaking, the state of a system is fully determined on an interval if it is given at the initial point of the interval. For a more complete discussion of the notion of *state*, the reader is referred to Ref. 9.1.

The control as well as the state variables may be not freely variable; that is, they may be subject to *constraints*. For instance, in the example discussed above, considerations of rocket engine design impose an upper bound on the thrust that can be generated. Thus the mass flow rate $\beta(\cdot)$ can vary between a minimum value, namely, zero corresponding to engine shut-down, and some maximum value β_{max} dictated by engine design; this bound may be constant or depend on the state. Note that no constraints were imposed in the formulation of this problem in Chapter 1; indeed, such constraints would place the problem outside the scope of the calculus of variations as treated in Part I. The state is freely variable in our example. In other problems, however, *state constraints* may arise. We shall begin with a discussion of systems involving only *control constraints*; subsequently we shall consider other kinds of constraints.

Thus far we have taken into account the ingredients which define the system under consideration: the independent variable, the control variables, and the state variables, together with the equations which determine the latter variables' evolution. That evolution is to be influenced by an appropriate choice of the control to accomplish a certain goal. This goal is twofold. First of all, the control function, here the mass flow rate $\beta(\cdot)$, must be such that, given the initial values of the state variables, $x(t_0) = x^0$, $v(t_0) = v^0$, and $m(t_0) = m^0$, the solution of the state equations (1.3) over some *unspecified* interval $[t_0, t_1]$ is such that $v(t_1) = v^1$ and $m(t_1) = m^1$, where v^1 and m^1 are prescribed; that is, the chosen control must "steer" the state from given initial to given terminal values of some of the state variables. In addition, it is desired to select an "optimal" control. In our example, an optimal $\beta(\cdot)$ should be such as to result in the maximum value of the range $x(t_1) - x(t_0) = \int_{t_0}^{t_1} v(t)\, dt$. In general, given a functional whose value depends on the control variables, one desires to extremize that value by appropriate choice of the control variables.

10

Problem Statement and Optimality

10.1. Introduction

This chapter is devoted to a discussion of what might be termed the "simplest problem of optimal control." As indicated in Chapter 9, it is the problem of extremizing the values of functionals which depend on control variables. These control variables belong to a given class of functions, satisfy constraints, and steer the state of the system under consideration from a prescribed initial state to a state that is wholly or partly specified.

In the following sections we give a general statement of the problems to be treated. In particular, we define various notions, especially that of optimal control, and present some basic properties of optimally controlled systems.

10.2. Problem Statement

Consider a system described by n *state variables* $x_1, x_2, \ldots, x_n$. Thus the system's state is a vector

$$x \triangleq \begin{bmatrix} x_1 \\ x_2 \\ \vdots \\ x_n \end{bmatrix},$$

or, geometrically speaking, it may be thought of as a point in an n-dimensional Euclidean space; $x \in R^n$.

The state may vary with respect to some scalar variable $t \in R^1$. That variable stands for time in many applications; however, it may be representative of other appropriate quantities such as arc length, angle, and so on.

Another set of variables entering into the system description are the *control variables* $u_1, u_2, \ldots, u_m$, which are the components of the control vector

$$u \triangleq \begin{bmatrix} u_1 \\ u_2 \\ \vdots \\ u_m \end{bmatrix};$$

that is, $u \in R^m$.

Now, given n functions

$$f_j(\cdot): R^n \times R^m \to R^1, \qquad j = 1, 2, \ldots, n,$$

and a control vector function

$$u(\cdot): [\bar{t}_0, \bar{t}_1] \to R^m,$$

the state evolution over $[t_0, t_1] \subset [\bar{t}_0, \bar{t}_1]$, namely,

$$x(\cdot): [t_0, t_1] \to R^n, \qquad x(t_0) = x^0,$$

must satisfy the *state equations*

$$\dot{x}_j(t) = f_j[x(t), u(t)], \qquad j = 1, 2, \ldots, n. \tag{10.1}$$

Letting

$$f(x, u) \triangleq \begin{bmatrix} f_1(x, u) \\ f_2(x, u) \\ \vdots \\ f_n(x, u) \end{bmatrix},$$

we may write (10.1) in vector notation as

$$\dot{x}(t) = f[x(t), u(t)]. \tag{10.2}$$

Chap. 10 • Problem Statement and Optimality

In order to assure that the state equations possess a unique solution for a given initial condition, we introduce certain hypotheses. Henceforth we assume that the functions $f_j(\cdot)$, $j=1,2,\ldots,n$, as well as their partial derivatives with respect to the state vector components, are continuous functions of the state and control; namely, we suppose that $f(\cdot): R^n \times R^m \to R^n$ is such that $f_j(\cdot)$ and $\partial f_j(\cdot)/\partial x_i$, $i,j=1,2,\ldots,n$, are defined and continuous on $R^n \times R^m$.

Next we impose conditions on the control variables. A control function $u(\cdot):[\bar{t}_0, \bar{t}_1] \to R^m$ is *admissible* if and only if (i) it is defined and piecewise continuous, and (ii) $u(t) \in U$ for all $t \in [\bar{t}_0, \bar{t}_1]$, where $U \subset R^m$ is a prescribed constraint set.

The use of admissible controls assures now that given an initial state $x(t_0) = x^0$, $t_0 \in [\bar{t}_0, \bar{t}_1]$, the state equation (10.2) possesses a unique piecewise smooth solution $x(\cdot):[t_0, t_1] \to R^n$, $x(t_0) = x^0$, where $[t_0, t_1] \subset [\bar{t}_0, \bar{t}_1]$; for instance, see Ref. 10.1.

Note that the independent variable t does not enter explicitly in the right-hand side of (10.2). We call such a state equation *autonomous*. If an admissible control $u(\cdot):[\bar{t}_0, \bar{t}_1] \to R^m$ leads to, or, as we shall say, *generates* a solution $x(\cdot):[t_0, t_1] \to R^n$, $x(t_0) = x^0$, of (10.2), then the control function $\bar{u}(\cdot):[\bar{t}_0 + \Delta t, \bar{t}_1 + \Delta t] \to R^m$ such that $\bar{u}(t + \Delta t) = u(t)$ leads to the solution $\bar{x}(\cdot):[t_0 + \Delta t, t_1 + \Delta t] \to R^n$, $\bar{x}(t_0 + \Delta t) = x^0$, such that $\bar{x}(t + \Delta t) = x(t)$. This is a trivial translation by Δt of the intervals of definition of $u(\cdot)$ and $x(\cdot)$.

For the present we are concerned with problems in which the *initial state is prescribed* and the *terminal state belongs to a given subset*, θ^1, of state space R^n; we refer to θ^1 as the *target* set. Thus we require that

$$\begin{aligned} x(t_0) &= x^0 \quad \text{(given)}, \\ x(t_1) &\in \theta^1 \quad \text{(given)}. \end{aligned} \tag{10.3}$$

Subsequently we shall impose additional conditions on the target set θ^1; for the time being we consider it to be simply a prescribed nonempty subset of R^n.

Since we are concerned only with solutions of (10.2) which satisfy the end conditions (10.3), we restrict consideration to those admissible controls which generate solutions emanating from x^0 and terminating on θ^1. A control $u(\cdot):[t_0, t_1] \to R^m$ is *feasible at* x^0 if and only if it is admissible and generates the solution $x(\cdot):[t_0, t_1] \to R^n$ such that $x(t_0) = x^0$ and $x(t_1) \in \theta^1$. We let $\mathcal{U}(x^0)$ denote the set of all controls which are feasible at x^0.

Now, given a function $f_0(\cdot): R^n \times R^m \to R^1$ of the same class as the functions $f_j(\cdot)$, $j = 1, 2, \ldots, n$, we introduce a *cost*—also called a criterion or performance index—given by

$$V[x^0, u(\cdot), x(\cdot)] \triangleq \int_{t_0}^{t_1} f_0[x(t), u(t)] \, dt \qquad (10.4)$$

where $u(\cdot):[t_0, t_1] \to R^m$ is an admissible control that generates the solution $x(\cdot):[t_0, t_1] \to R^n$, $x(t_0) = x^0$, $[t_0, t_1] \subset [\bar{t}_0, \bar{t}_1]$. The value of the cost depends on the control $u(\cdot)$ and on the solution $x(\cdot)$ generated by it. Thus the cost is a functional.

Note that the choice of t_0 does not affect the value of the cost; namely,

$$\int_{t_0}^{t_1} f_0[x(t), u(t)] \, dt = \int_{t_0 + \Delta t}^{t_1 + \Delta t} f_0[\bar{x}(t), \bar{u}(t)] \, dt.$$

We are now ready to state the *simplest problem of optimal control*: Given the state equation (10.2), the control constraint set U, the target set θ^1, and the cost (10.4), find a control in $\mathcal{U}(x^0)$ that minimizes the value of the cost over all controls in $\mathcal{U}(x^0)$. Thus, $u^*(\cdot):[t_0^*, t_1^*] \to R^m$, generating the solution $x^*(\cdot):[t_0^*, t_1^*] \to R^n$ such that $x^*(t_0^*) = x^0$ and $x^*(t_1^*) \in \theta^1$, is *optimal at x^0* if and only if

$$V[x^0, u^*(\cdot), x^*(\cdot)] \leq V[x^0, u(\cdot), x(\cdot)] \qquad (10.5)$$

for all $u(\cdot) \in \mathcal{U}(x^0)$, that is, for every admissible $u(\cdot):[t_0, t_1] \to R^m$ generating the solution $x(\cdot):[t_0, t_1] \to R^n$, $x(t_0) = x^0$ and $x(t_1) \in \theta^1$.

The state equations are autonomous, the target set is a constant subset of R^n, and the cost integrand does not depend explicitly on t; that is, *the system is autonomous*. Then, in view of the trivial translation property of the interval $[t_0, t_1]$, no loss of generality is caused by fixing the value of t_0—that is, $t_0 = t_0^* =$ constant—and considering admissible $u(\cdot)$ defined on all compact intervals $[t_0, t_1]$. Also, as already mentioned in Part I, defining optimality in terms of minimizing does not preclude maximizing, since a control that minimizes the value of $-V(\cdot)$ maximizes the value of $V(\cdot)$.

Henceforth we take as *given*: the state equation (10.2), the control constraint set U, the target set θ^1, and the cost (10.4).

Consider an admissible control $u(\cdot):[t_0, t_1] \to R^m$ that generates the solution $x(\cdot):[t_0, t_1] \to R^n$, $x(t_0) = x^0$. The corresponding cost is given by (10.4); hence it follows that

$$\lim_{t \to t_1} V\big[x(t), u(\cdot)|_{[t, t_1]}, x(\cdot)|_{[t, t_1]}\big] = 0, \qquad t \in [t_0, t_1].$$

Thus, if $x^0 \in \theta^1$ one must consider the situation in which $t_1 = t_0$ and $x(t_1) = x(t_0) = x^0$; in this event the control is irrelevant and the corresponding cost is zero. Hence there are two possibilities:

(1) There is no control that is feasible at $x^0 \in \theta^1$ *and* results in a lower and hence *negative* cost; consequently, the minimum value of the cost is zero.
(2) There is an optimal control at $x^0 \in \theta^1$, say $u^*(\cdot) : [t_0, t_1^*] \to R^m$ with $t_1^* > t_0$; that is, the value of the cost can be lowered, or at least not increased, and hence must be nonpositive.

The subsequent discussion of optimal control at $x^0 \in \theta^1$ is concerned with the possibility (2). Of course, if $x^0 \notin \theta^1$, then $t_1^* > t_0$.

Finally it should be noted that some authors require that feasible controls generate solutions for which only the terminal value belongs to the target; that is, a control is feasible at x^0 if and only if the solution $x(\cdot) : [t_0, t_1] \to R^n$ generated by it satisfies (10.3) *and* $x(t) \notin \theta^1$ for $t < t_1$. We do not impose this restriction since it may preclude the existence of optimal controls; for instance, see Refs. 10.2 and 10.3.

10.3. Joining Controls and Additivity of Costs

The following properties are of fundamental importance to our subsequent discussion. Indeed, without them many results need not be valid.

Consider two admissible control functions, $u'(\cdot) : [t_0, t'] \to R^m$ and $u''(\cdot) : [t', t_1] \to R^m$. Consider also the function $u(\cdot) : [t_0, t_1] \to R^m$ such that

$$u(t) = u'(t) \quad \text{for } t \in [t_0, t'),$$
$$u(t) = u''(t) \quad \text{for } t \in [t', t_1].$$

In other words, the control $u(\cdot)$ is defined by "joining" $u'(\cdot)$ and $u''(\cdot)$; see Figure 10.1. Then $u(\cdot)$ is an admissible control since it satisfies the conditions (i) and (ii) of the definition of an admissible control; namely, $u(\cdot)$ is piecewise continuous and $u(t) \in U$ for all $t \in [t_0, t_1]$. Note that one could not assure this property if one were to restrict admissible controls to the class of *continuous* functions.

Now consider again two admissible controls, $u'(\cdot) : [t_0, t'] \to R^m$ generating the solution $x'(\cdot) : [t_0, t'] \to R^m$, $x'(t_0) = x^0$, and $u''(\cdot) : [t', t_1] \to R^m$ generating the solution $x''(\cdot) : [t', t_1] \to R^n$, $x''(t') = x'(t')$; see Figure 10.2. Then the control $u(\cdot) : [t_0, t_1] \to R^m$, defined by joining $u'(\cdot)$ and $u''(\cdot)$, generates

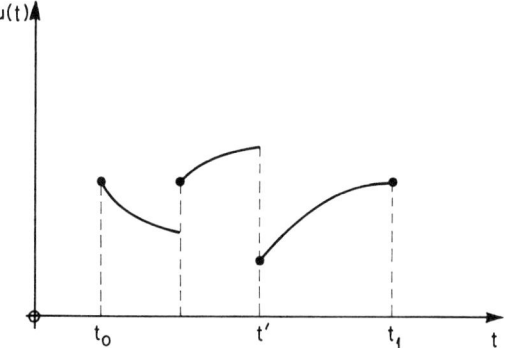

Figure 10.1. Joining controls.

the solution $x(\cdot):[t_0, t_1] \to R^n$, $x(t_0)=x^0$, where

$$x(t) = x'(t) \quad \text{for } t \in [t_0, t'],$$
$$x(t) = x''(t) \quad \text{for } t \in [t', t_1].$$

Consider also the corresponding costs

$$V[x^0, u'(\cdot), x'(\cdot)] = \int_{t_0}^{t'} f_0[x'(t), u'(t)]\, dt,$$

$$V[x'(t'), u''(\cdot), x''(\cdot)] = \int_{t'}^{t_1} f_0[x''(t), u''(t)]\, dt,$$

$$V[x^0, u(\cdot), x(\cdot)] = \int_{t_0}^{t_1} f_0[x(t), u(t)]\, dt.$$

Then it follows at once that

$$V[x^0, u'(\cdot), x'(\cdot)] + V[x'(t'), u''(\cdot), x''(\cdot)] = V[x^0, u(\cdot), x(\cdot)].$$
(10.6)

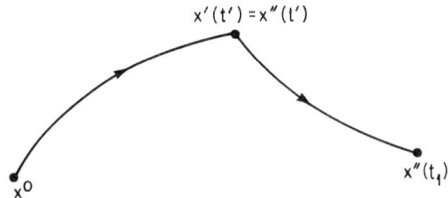

Figure 10.2. A solution generated by joined controls.

Chap. 10 • Problem Statement and Optimality

In other words, the total cost along the solution $x(\cdot)$ is the sum of the costs along $x'(\cdot)$ and $x''(\cdot)$, respectively; that is, *costs are additive.*

As already noted at the end of the last section, another immediate consequence of the definition (10.4) of the cost is that

$$\lim_{t \to t_1} V\big[x(t), u(\cdot)|_{[t,t_1]}, x(\cdot)|_{[t,t_1]}\big] = \lim_{t \to t_1} \int_t^{t_1} f_0[x(\tau), u(\tau)] \, d\tau$$
$$= 0. \tag{10.7}$$

10.4. Optimal Cost and an Optimality Principle

Before deducing some basic properties associated with an optimal control, let us introduce a few more definitions.

Consider two subsets of the state space R^n. Let

$$E \triangleq \{x^0 \in R^n | \text{ there exists a feasible control at } x^0\}; \tag{10.8}$$

that is, E is the set of all states which can be steered to the target set θ^1. Thus no state in $R^n \setminus E$ can be steered to θ^1.

Let

$$E^* \triangleq \{x^0 \in R^n | \text{ there exists an optimal control at } x^0\}. \tag{10.9}$$

Since optimality presupposes feasibility, $E^* \subset E$. Henceforth it is assumed that E^* is not empty.

Now let $u^*(\cdot).[t_0, t_1^*] \to R^m$ be an optimal control at x^0 that generates the solution $x^*(\cdot):[t_0, t_1^*] \to R^n$, $x^*(t_0) = x^0$. Then we define the *optimal cost function*, also sometimes called the optimal return function (for instance, see Refs. 10.4–10.6), $V^*(\cdot): E^* \to R^1$ such that for $x^0 \in E^*$

$$V^*(x^0) \triangleq V[x^0, u^*(\cdot), x^*(\cdot)]. \tag{10.10}$$

For a given initial state x^0 there may exist more than one optimal control; that is, the optimal control at x^0 need not be unique. However, by definition of a minimum, provided it exists, $V^*(x^0)$ is unique; hence, $V^*(\cdot)$ is indeed a function from E^* into R^1.

We are now ready to deduce a property of optimal control often referred to as a *principle of optimality*; for instance, see Refs. 10.4–10.6.

Lemma 10.1. *Let* $u^*(\cdot):[t_0, t_1^*] \to R^m$ *be an optimal control at* x^0 *that generates the solution* $x^*(\cdot):[t_0, t_1^*] \to R^n$, $x^*(t_0) = x^0$. *Then the restriction* $u^*(\cdot)|_{[\bar{t}, t_1^*]}$ *is an optimal control at* $x^*(\bar{t})$, $\bar{t} \in [t_0, t_1^*)$.

Proof. By the definition of the optimal cost function (10.10) and by the cost additivity (10.6), we have

$$V^*(x^0) \triangleq V[x^0, u^*(\cdot), x^*(\cdot)]$$

$$= V[x^0, u'(\cdot), x'(\cdot)] + V[x^*(\bar{t}), u''(\cdot), x''(\cdot)] \quad (10.11)$$

where (see Figure 10.3)

$$u'(\cdot) = u^*(\cdot)|_{[t_0, \bar{t}]}, \quad x'(\cdot) = x^*(\cdot)|_{[t_0, \bar{t}]},$$

$$u''(\cdot) = u^*(\cdot)|_{[\bar{t}, t_1^*]}, \quad x''(\cdot) = x^*(\cdot)|_{[\bar{t}, t_1^*]}$$

if $u^*(\cdot)$ is continuous at $t = \bar{t}$. If $u^*(\cdot)$ is discontinuous at $t = \bar{t}$, let

$$u'(\bar{t}) = \lim_{\substack{t \to \bar{t} \\ t < \bar{t}}} u^*(t) \triangleq u^*(\bar{t} - 0),$$

$$u''(\bar{t}) = \lim_{\substack{t \to \bar{t} \\ t > \bar{t}}} u^*(t) \triangleq u^*(\bar{t} + 0).$$

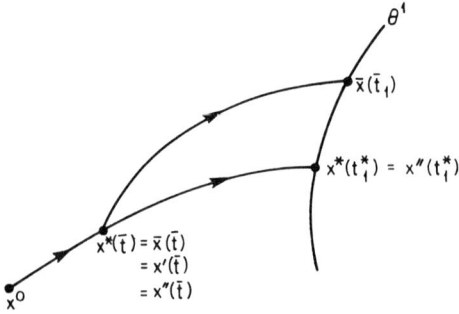

Figure 10.3. Solutions in the proof of Lemma 10.1.

Chap. 10 • Problem Statement and Optimality

Now suppose that $u''(\cdot)$ is not optimal at $x^*(\bar{t})$. Then there exists a feasible control at $x^*(\bar{t})$, say $\bar{u}(\cdot) : [\bar{t}, \bar{t}_1] \to R^m$ generating the solution $\bar{x}(\cdot): [\bar{t}, \bar{t}_1] \to R^n$, $\bar{x}(\bar{t}) = x^*(\bar{t})$ and $\bar{x}(\bar{t}_1) \in \theta^1$, such that

$$V[x^*(\bar{t}), \bar{u}(\cdot), \bar{x}(\cdot)] < V[x^*(\bar{t}), u''(\cdot), x''(\cdot)]. \tag{10.12}$$

By joining $u'(\cdot)$ and $\bar{u}(\cdot)$ one obtains an admissible control—indeed, one that is feasible at x^0—say $u(\cdot):[t_0, \bar{t}_1] \to R^m$ generating the solution $x(\cdot):[t_0, \bar{t}_1] \to R^n$, $x(t_0) = x^0$, where

$$x(\cdot)|_{[t_0, \bar{t}]} = x'(\cdot), \qquad x(\cdot)|_{[\bar{t}, \bar{t}_1]} = \bar{x}(\cdot).$$

Thus, in view of the cost additivity together with (10.11) and (10.12), we have

$$V[x^0, u(\cdot), x(\cdot)] < V^*(x^0).$$

But this contradicts the hypothesized optimality of $u^*(\cdot)$ at x^0. Consequently, $u''(\cdot) = u^*(\cdot)|_{[\bar{t}, t_1^*]}$ is optimal at $x^*(\bar{t})$, $\bar{t} \in [t_0, t_1^*)$. □

Consider

$$V^*(x^*(\bar{t})) = V\left[x^*(\bar{t}), u^*(\cdot)|_{[\bar{t}, t_1^*]}, x^*(\cdot)|_{[\bar{t}, t_1^*]}\right].$$

In view of the definition (10.7), it follows that

$$\lim_{\bar{t} \to t_1^*} V^*(x^*(\bar{t})) = 0. \tag{10.13}$$

Now let θ^* be the set of terminal values of the solutions generated by *all* optimal controls at *every* initial state; that is,

$$\theta^* \triangleq \{x \in \theta^1 | x = x^*(t_1^*), x^0 \in E^*\}. \tag{10.14}$$

Then we may extend the domain of definition of $V^*(\cdot)$ to $E^* \cup \theta^*$, with

$$V^*(x) = 0 \qquad \forall x \in \theta^* \tag{10.15}$$

in order to satisfy (10.13). Of course, $V^*(x) = 0$ for $x \in E^* \cap \theta^*$; see Exercise 10.5.

Now it is an immediate consequence of Lemma 10.1 that $x^*(t) \in E^* \cup \theta^*$ for all $t \in [t_0, t_1^*]$.

10.5. Augmented State Space and Trajectories

In preparation for a discussion of geometric properties associated with optimal control, some additional quantities must be defined.

We begin by augmenting the state space R^n by another dimension corresponding to the cost. The *augmented state* is denoted by

$$y \triangleq [x_0 \ x_1 \ x_2 \cdots x_n]^T \in R^{n+1}. \tag{10.16}$$

Then, in the augmented state space R^{n+1}, we define the sets

$$\Theta^1 \triangleq R^1 \times \theta^1, \qquad \Theta^* \triangleq R^1 \times \theta^*,$$

$$\mathcal{E} \triangleq R^1 \times E, \qquad \mathcal{E}^* \triangleq R^1 \times E^*. \tag{10.17}$$

Now given an admissible control $u(\cdot):[t_0, t_1] \to R^m$ with the corresponding solution $x(\cdot):[t_0, t_1] \to R^n$, $x(t_0) = x^0$, we introduce the *cost variable* $x_0(\cdot):[t_0, t_1] \to R^1$, $x_0(t_1) = C$, such that

$$x_0(t) = C - \int_t^{t_1} f_0[x(\tau), u(\tau)] \, d\tau$$

$$= C - V\Big[x(t), u(\cdot)|_{[t,t_1]}, x(\cdot)|_{[t,t_1]}\Big]. \tag{10.18}$$

Since

$$x_0^0 + V[x^0, u(\cdot), x(\cdot)] = C, \qquad x_0(t_0) = x_0^0, \tag{10.19}$$

we *may* think of x_0^0 as the cost of transfer to the initial state x^0 and of C as the *total* cost of transfer to $x(t_1)$. However, the quantity of interest to us is the cost of transfer from x^0 to $x(t_1)$, namely, $C - x_0^0$. Thus either C or x_0^0 is arbitrary.

In view of (10.18) and the state equation (10.2) we can introduce an *augmented state equation*. Let $h(\cdot): R^{n+1} \times R^m \to R^{n+1}$ be such that for $i = 0, 1, \ldots, n$

$$h_i(y, u) = f_i(x, u) \qquad \forall y \in R^{n+1} \text{ and } u \in R^m. \tag{10.20}$$

Then $y(\cdot):[t_0, t_1] \to R^{n+1}$, $y(t_0) = y^0$, is a solution of

$$\dot{y}(t) = h[y(t), u(t)]. \tag{10.21}$$

Chap. 10 • Problem Statement and Optimality

Given a control $u(\cdot)$ generating the solution $x(\cdot)$ of (10.2), a corresponding *trajectory* in R^n is the ordered set of points

$$\{x \in R^n | x = x(t), t \in [t_0, t_1]\}.$$

Then, given also the value of the constant C in (10.18), a corresponding *trajectory* in R^{n+1} is the ordered set of points

$$\Gamma(C) \triangleq \{y \in R^{n+1} | y = y(t), t \in [t_0, t_1]\}. \tag{10.22}$$

Henceforth it should be clear from the context whether we refer to a solution $x(\cdot)$ of the state equation and a corresponding trajectory in R^n or to a solution $y(\cdot)$ of the augmented state equation and a corresponding trajectory in R^{n+1}.

By varying the value of the parameter C one generates a one-parameter family of trajectories in R^{n+1} which lie in the x_0-cylindrical surface whose intersection with $\{y \in R^{n+1} | x_0 = 0\}$ is the trajectory in R^n corresponding to the solution $x(\cdot)$; see Figure 10.4.

If $u(\cdot) \in \mathcal{U}(x^0)$ so that $x^0 \in E$ and $x(t_1) \in \theta^1$, and $y^0 \in \mathcal{E}$ and $y(t_1) \in \Theta^1$, then $\Gamma(C)$ is a *terminating trajectory*.

If $u^*(\cdot) : [t_0, t_1^*] \to R^m$ is optimal at $x^0 \in E^*$ and generates the solution $x^*(\cdot) : [t_0, t_1^*] \to R^n$, $x^*(t_0) = x^0$, then we define an *optimal trajectory* for a

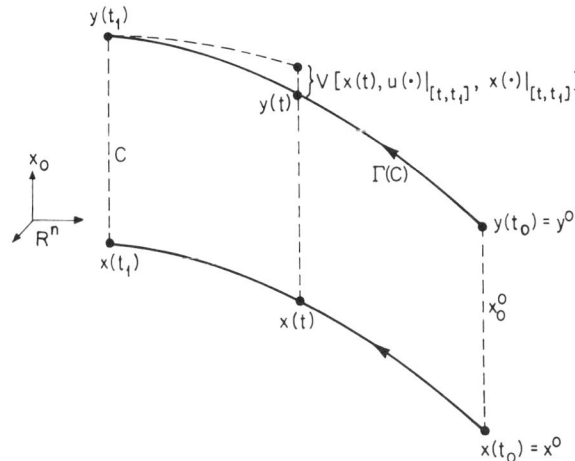

Figure 10.4. Trajectories.

given C by

$$\Gamma^*(C) \triangleq \{y \in R^{n+1} | y = y^*(t), t \in [t_0, t_1^*]\} \qquad (10.23)$$

where $y^*(t) \triangleq [x_0^*(t)\ x_1^*(t) \cdots x_n^*(t)]^T$ with $x(\tau) = x^*(\tau)$ and $u(\tau) = u^*(\tau)$ in (10.18). Of course, $\Gamma^*(C)$ is terminating, $y^0 \in \mathcal{E}^*$ and $y^*(t_1^*) \in \Theta^*$. Varying C again leads to a one-parameter family of such optimal trajectories in R^{n+1}.

Shortly we shall describe the behavior of trajectories and of optimal trajectories with respect to certain "surfaces" in R^{n+1}.

10.6. Limiting and Optimal Isocost Surfaces

The optimal cost function $V^*(\cdot): E^* \cup \theta^* \to R^1$ is defined in Section 10.4. By means of this function we define now two kinds of "surfaces" in R^{n+1} and in R^n, respectively. The term "surface" as used here refers merely to a set of points defined by an equation; no other properties are implied.

Given a constant C, we define a *limiting surface* by

$$\Sigma(C) \triangleq \{y \in R^{n+1} | x_0 = C - V^*(x), x \in E^* \cup \theta^*\}. \qquad (10.24)$$

Such a surface is single-sheeted; that is, its points are in one-to-one correspondence with the points of $E^* \cup \theta^*$. One and only one such surface passes through a given point of $\mathcal{E}^* \cup \Theta^*$. For consider two surfaces, $\Sigma(C_1)$ and $\Sigma(C_2)$, $C_1 \neq C_2$, and let

$$y^1 = \begin{bmatrix} x_0^1 \\ -- \\ x \end{bmatrix} \in \Sigma(C_1), \quad y^2 = \begin{bmatrix} x_0^2 \\ -- \\ x \end{bmatrix} \in \Sigma(C_2).$$

Then, in view of the definition (10.24),

$$x_0^1 - x_0^2 = C_1 - C_2 \quad \text{for all } x \in E^* \cup \theta^*.$$

Thus the one-parameter family of limiting surfaces, generated by varying C, consists of surfaces deduced from one another by x_0-parallel translation; see Figure 10.5.

Another one-parameter family of surfaces, now in R^n, consists of the so-called *optimal isocost surfaces* defined by

$$S(C) \triangleq \{x \in R^n | V^*(x) = C, x \in E^* \cup \theta^*\}. \qquad (10.25)$$

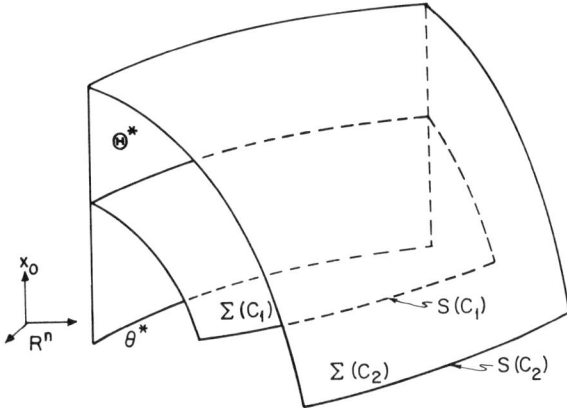

Figure 10.5. Limiting surfaces.

Note that (see Figure 10.5)

$$\Sigma(C) \cap \{y \in R^{n+1} | x_0 = 0\} = \{y \in R^{n+1} | x_0 = 0, x \in S(C)\}.$$

In the trivial case of the optimal cost being the same for all initial states, namely,

$$V^*(x) = 0 \qquad \forall x \in E^* \cup \theta^*$$

we have

$$S(0) = E^* \cup \theta^*$$

and

$$\Sigma(C) \cap \{y \in R^{n+1} | x_0 = 0\} = \emptyset \qquad \text{for } C \neq 0.$$

The name "limiting surface" is used because of a property to be derived in the next section. The appellation "optimal isocost surface" is a consequence of the definitions (10.10) and (10.25); namely, $S(C)$ is the locus of all initial states from which the state can be transferred to the target θ^1 with the *same minimum cost*, C.

Before discussing the behavior of trajectories, both optimal and nonoptimal, in relation to limiting surfaces, we require another definition. A given limiting surface, $\Sigma(C)$, separates $\mathcal{E}^* \cup \Theta^*$ into two disjoint regions in R^{n+1},

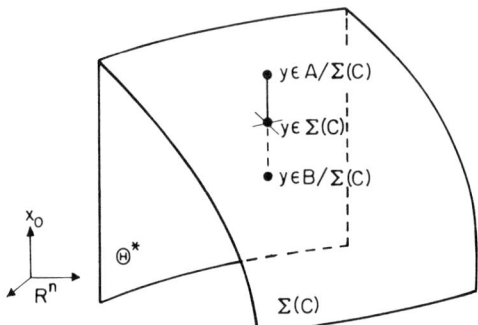

Figure 10.6. The regions of $\mathscr{E}^* \cup \Theta^*$.

namely, the set of points "above" $\Sigma(C)$ and the set of points "below" $\Sigma(C)$. We denote these regions by

$$A/\Sigma(C) \triangleq \{y \in R^{n+1} | x_0 > C - V^*(x), x \in E^* \cup \theta^*\},$$
$$B/\Sigma(C) \triangleq \{y \in R^{n+1} | x_0 < C - V^*(x), x \in E^* \cup \theta^*\} \quad (10.26)$$

where

$$y = \begin{bmatrix} x_0 \\ -- \\ x \end{bmatrix}.$$

Thus we have

$$\mathscr{E}^* \cup \Theta^* = A/\Sigma(C) \cup \Sigma(C) \cup B/\Sigma(C).$$

This is illustrated in Figure 10.6. If $y \in A/\Sigma(C)$ [$B/\Sigma(C)$] it is termed an A-point [B-point] relative to $\Sigma(C)$.

10.7. Fundamental Properties of Trajectories and of Limiting Surfaces

In this section we deduce fundamental properties of trajectories vis-à-vis limiting surfaces. In particular we show that a limiting surface is the locus of all optimal trajectories whose initial points belong to that surface. Furthermore a nonoptimal trajectory whose initial point lies on a particular limiting surface cannot dip "below" that surface. We have the following theorem.

Chap. 10 • Problem Statement and Optimality

Theorem 10.1. *If $\Gamma^*(C)$ is an optimal trajectory whose initial point belongs to the limiting surface $\Sigma(\bar{C})$, then $\bar{C}=C$ and $\Gamma^*(C)$ lies entirely in $\Sigma(C)$. If $\Gamma(\tilde{C})$ is a nonoptimal trajectory whose initial point belongs to the limiting surface $\Sigma(C)$, then no point of $\Gamma(\tilde{C})$ is a B-point relative to $\Sigma(C)$.*

Proof. Consider the first part of the theorem. Let $u^*(\cdot):[t_0, t_1^*]\to R^m$ be the optimal control at x^0 that generates the solution $y^*(\cdot):[t_0, t_1^*]\to R^{n+1}$, where

$$y^*(t_0)=\begin{bmatrix} x_0^0 \\ \overline{x^0} \end{bmatrix} \in \Sigma(\bar{C}),$$

corresponding to the trajectory $\Gamma^*(C)$. Since $y^*(t_0)\in\Sigma(\bar{C})$, it follows from (10.23) and (10.24) with (10.10) that $\bar{C}=C$; hence, $y^*(t_0)\in\Sigma(\bar{C})\equiv\Sigma(C)$.

Now consider

$$y^*(t)=\begin{bmatrix} x_0^*(t) \\ \overline{x^*(t)} \end{bmatrix}, \quad t\in[t_0, t_1^*].$$

By (10.23) we have

$$x_0^*(t)+V\big[x^*(t), u^*(\cdot)|_{[t, t_1^*]}, x^*(\cdot)|_{[t, t_1^*]}\big]=C,$$

and by (10.24) with $\bar{C}=C$

$$x_0+V^*(x^*(t))=C$$

so that, in view of Lemma 10.1 and (10.10),

$$x_0=x_0^*(t),$$

and hence $y^*(t)\in\Sigma(C)$.

Next we turn to the second part of the theorem. Let $u(\cdot):[t_0, t_1]\to R^m$ denote the admissible control that generates the solution $y(\cdot):[t_0, t_1]\to R^{n+1}$ corresponding to the trajectory $\Gamma(\tilde{C})$, where

$$y(t_0)=\begin{bmatrix} x_0(t_0) \\ \overline{x(t_0)} \end{bmatrix}\in\Sigma(C).$$

Now suppose that

$$y(\hat{t}) = \begin{bmatrix} x_0(\hat{t}) \\ \hline x(\hat{t}) \end{bmatrix}, \quad \hat{t} \in [t_0, t_1],$$

is a B-point relative to $\Sigma(C)$ so that $y(\hat{t}) \in \mathcal{E}^* \cup \Theta^*$. Then it follows that

$$x_0(t_0) + V^*(x(t_0)) = C \tag{10.27}$$

and

$$x_0(\hat{t}) + V^*(x(\hat{t})) = \bar{C} < C. \tag{10.28}$$

Also, by (10.22), we have

$$x_0(\hat{t}) = x_0(t_0) + \int_{t_0}^{\hat{t}} f_0[x(t), u(t)] \, dt. \tag{10.29}$$

Upon substituting for $x_0(\hat{t})$ from (10.29) in (10.28) and employing (10.27), we obtain

$$\int_{t_0}^{\hat{t}} f_0[x(t), u(t)] \, dt + V^*(x(\hat{t})) < V^*(x(t_0)). \tag{10.30}$$

Now either $y(\hat{t}) \in \Theta^*$ or $y(\hat{t}) \notin \Theta^*$. In the former eventuality, $V^*(x(\hat{t})) = 0$ by (10.15). But then $u(\cdot)|_{[t_0, \hat{t}]}$ is feasible at $x(t_0)$ *and* results in a cost that is less than $V^*(x(t_0))$. That contradicts the optimality of $V^*(x(t_0))$. In the latter eventuality, $y(\hat{t}) \notin \Theta^*$ but $y(\hat{t}) \in \mathcal{E}^*$. Thus there exists an optimal

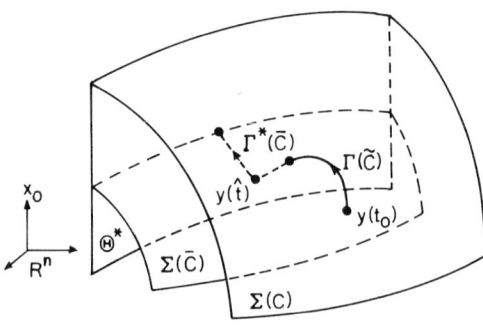

Figure 10.7. A union of trajectories in the proof of Theorem 10.1.

trajectory with the initial point $y(\hat{t}) \in \Sigma(\overline{C})$, say $\Gamma^*(\overline{C}) \subset \Sigma(\overline{C})$. Now consider the union of the portion of $\Gamma(\tilde{C})$ for $t \in [t_0, \hat{t}]$ with $\Gamma^*(\overline{C})$; see Figure 10.7. This trajectory is generated by a control that is feasible at $x(t_0)$ and results in the cost

$$\int_{t_0}^{\hat{t}} f_0[x(t), u(t)] \, dt + V^*(x(\hat{t})).$$

Thus, as a consequence of (10.30), the optimality of $V^*(x(t_0))$ is contradicted. □

Loosely speaking, according to Theorem 10.1, all trajectories—optimal or not—which start on a given limiting surface lie entirely on or "above" that surface; the surface "limits" the region in which such trajectories lie.

10.8. An Illustrative Example

Consider a system characterized by a single state variable that evolves according to the *state equation*

$$\dot{x}_1(t) = u(t), \qquad (10.31)$$

with *control constraint set*

$$U = \{ u \in R^1 \, | \, |u| \leq 1 \}. \qquad (10.32)$$

The state is to be transferred from a given initial state $x_1(t_0) = x_1^0$ to a given terminal state $x_1(t_1) = 0$ within an *unspecified* bounded interval $[t_0, t_1]$; that is, the target set θ^1 is simply the point $x = 0$ in the state space R^1.

The *cost* to be minimized is

$$V[x^0, u(\cdot), x(\cdot)] = \int_{t_0}^{t_1} dt = t_1 - t_0. \qquad (10.33)$$

Such a problem is termed "time-optimal" if t stands for time.

Now it is readily shown that $E = R^1$. For consider $u(t) \equiv \text{const} \triangleq c \neq 0, |c| \leq 1$. Then, by (10.31) and the given end conditions

$$t_1 - t_0 = -x_1^0/c.$$

Thus, given an $x_1^0, |x_1^0| < \infty$, there is a constant $c, |c| \leq 1$, for which $t_1 - t_0 < \infty$. For example, if $c = -k \, \text{sgn} \, x_1^0, k \in (0, 1]$, then $t_1 - t_0 = |x_1^0|/k$.

By employing (10.31) in (10.33) one obtains

$$t_1 - t_0 = \int_{x_1^0}^{0} \frac{dx_1(t)}{u(t)}.$$

It follows at once that the *optimal control* is given by

$$u^*(t) \equiv -\operatorname{sgn} x_1^0, \qquad (10.34)$$

clearly showing the dependence of $u^*(\cdot)$ on the initial state. The corresponding *minimum cost* is

$$t_1^* - t_0 = |x_1^0|. \qquad (10.35)$$

Since $u^*(\cdot)$ exists for all $x_1^0 \in E \setminus \{0\}$, $E^* \cup \theta^* = E$. Of course, $\theta^* = \theta^1 = \{0\}$.

The state equation of the *augmented state* is

$$\dot{y}(t) = \begin{bmatrix} \dot{x}_0(t) \\ \dot{x}_1(t) \end{bmatrix} = \begin{bmatrix} 1 \\ u(t) \end{bmatrix} \qquad (10.36)$$

so that an optimal trajectory is given by

$$\begin{aligned} x_0^*(t) &= x_0^0 + (t - t_0), \\ x_1^*(t) &= x_1^0 - (t - t_0) \operatorname{sgn} x_1^0. \end{aligned} \qquad (10.37)$$

Now, in view of (10.35), the *optimal cost* function is defined by

$$V^*(x) = |x_1| \qquad \forall x_1 \in R^1.$$

Thus the equation of a *limiting surface* $\Sigma(C)$ is

$$x_0 + |x_1| = C,$$

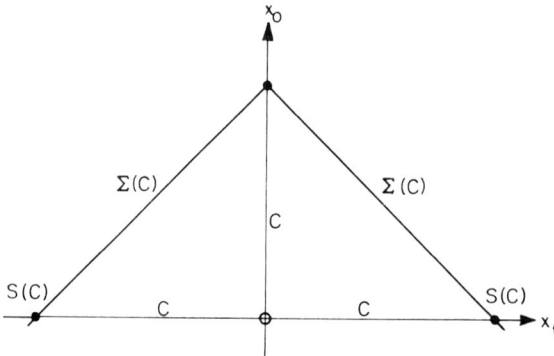

Figure 10.8. Limiting and optimal isocost surfaces.

Chap. 10 • Problem Statement and Optimality

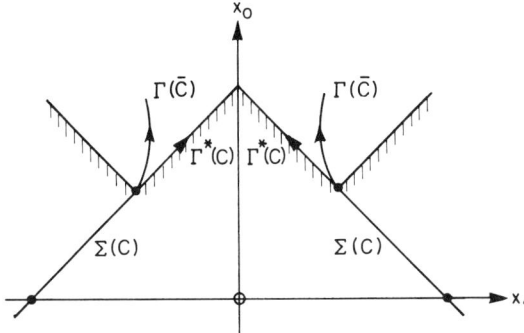

Figure 10.9. Properties of trajectories and limiting surfaces.

while that of an optimal isocost surface $S(C)$ is

$$|x_1| = C.$$

This is shown in Figure 10.8.

Finally let us verify the properties of $\Sigma(C)$ embodied in Theorem 10.1. As a consequence of (10.37) we have

$$\frac{x_0^*(t) - x_0^0}{x_1^*(t) - x_1^0} = \frac{-1}{\operatorname{sgn} x_1^0}.$$

Thus

$$y^*(t_0) = \begin{bmatrix} x_0^0 \\ x_1^0 \end{bmatrix} \in \Sigma(C)$$

implies that $y^*(t) \in \Sigma(C)$ for all $t \in [t_0, t_1^*]$. Moreover it follows from (10.36) that

$$\frac{dx_1(t)}{dx_0(t)} = u(t)$$

and thence, in view of the constraint (10.32), that

$$-1 \le \frac{dx_1(t)}{dx_0(t)} \le 1$$

so that, for $y^0 \in \Sigma(C)$, all trajectories belong to $A/\Sigma(C) \cup \Sigma(C)$; this is shown in Figure 10.9.

Exercises

10.1. Consider the system characterized by a single state variable evolving according to

$$\dot{x}(t) = u^2(t),$$

where control is subject to the constraint

$$|u(t)| \leq 1.$$

The state is to be transferred from the given initial state $x(t_0) = x^0$ to the given terminal state $x(t_1) = 0$ within an unspecified bounded interval $t_1 - t_0$. Deduce the set E of initial states for which there is a feasible control.

10.2. Consider the system of Exercise 10.1 and suppose that it is desired to minimize

$$\int_{t_0}^{t_1} dt = t_1 - t_0.$$

(a) Deduce an optimal control by inspection.
(b) What is the minimum cost for given x^0?
(c) Give the optimal cost function.

10.3. For the problem of Exercise 10.2, give
   (a) the augmented state equation;
   (b) the solution defining an optimal trajectory;
   (c) the set E^* of initial states for which there is an optimal control;
   (d) the equations for the limiting and optimal isocost surfaces.

10.4. For the problem of Exercise 10.2, demonstrate the validity of Theorem 10.1.

10.5. Using the joining property of admissible controls and the additivity property of costs, show that $V^*(x) = 0$ for $x \in E^* \cap \theta^*$.

11

Regular Optimal Trajectories

11.1. Regular Interior Points of a Limiting Surface

In the ensuing discussion of optimal control we shall utilize the fundamental properties, established in Chapter 10, at points of an optimal trajectory $\Gamma^*(C)$ where the limiting surface $\Sigma(C)$ possesses a tangent plane.

Let $N(\delta, \bar{y})$ denote a δ-neighborhood of $\bar{y} \in R^{n+1}$; that is,

$$N(\delta, \bar{y}) \triangleq \{y \in R^{n+1} | \|y - \bar{y}\| < \delta, \delta > 0\}.$$

Also let $\Phi(\cdot): \mathcal{E}^* \cup \Theta^* \to R^1$ be the function defined by

$$\Phi(y) \triangleq x_0 + V^*(x)$$

so that (10.24) becomes

$$\Sigma(C) \triangleq \{y \in \mathcal{E}^* \cup \Theta^* | \Phi(y) = C\}.$$

We say that $\bar{y} \in \Sigma(C)$ is a *regular interior point* of $\Sigma(C)$ if and only if there is a $\delta > 0$ such that

(i) $N(\delta, \bar{y}) \subset \mathcal{E}^* \cup \Theta^*$;
(ii) $\Phi(\cdot)$ is continuous on $N(\delta, \bar{y})$;
(iii) there is an n-dimensional plane, $T_{\Sigma(C)}(\bar{y})$, containing $\bar{y}$ such that there is a one-to-one correspondence

$$y = \bar{y} + \Delta y + o(\|\Delta y\|) \tag{11.1}$$

between

$$y \in N(\delta, \bar{y}) \cap \Sigma(C)$$

and

$$\bar{y} + \Delta y \in N(\delta, \bar{y}) \cap T_{\Sigma(C)}(\bar{y})$$

where $o(\|\Delta y\|)$ is normal to $T_{\Sigma(C)}(\bar{y})$ and where

$$\lim_{\|\Delta y\| \to 0} \frac{o(\|\Delta y\|)}{\|\Delta y\|} = 0.$$

Figure 11.1 is an illustration of a regular interior point $\bar{y}$ of $\Sigma(C)$.

Before proceeding with a derivation of necessary conditions which are met at a regular interior point of a limiting surface, we deduce some properties of tangent plane $T_{\Sigma(C)}(\bar{y})$. Consider a unit vector $n(\bar{y})$ at $\bar{y}$ that is normal to $T_{\Sigma(C)}(\bar{y})$. We assert that one can select that normal, of the two possible ones, which "points into $B/\Sigma(C)$"; namely, there is a $\gamma > 0$ such that

$$\bar{y} + \beta n(\bar{y}) \in B/\Sigma(C) \qquad \forall \beta \in (0, \gamma]. \tag{11.2}$$

We begin by showing that there is a $\gamma > 0$ such that *either*

$$\bar{y} + \beta n(\bar{y}) \in A/\Sigma(C) \qquad \forall \beta \in (0, \gamma]$$

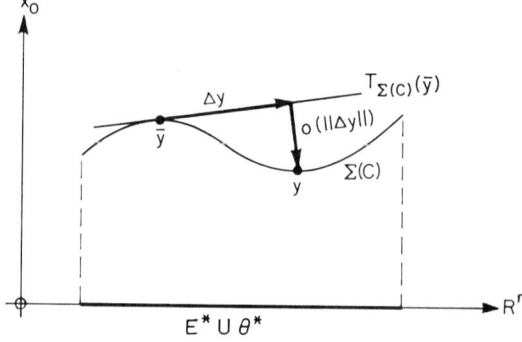

Figure 11.1. A regular interior point of a limiting surface.

or
$$\bar{y}+\beta n(\bar{y})\in B/\Sigma(C) \quad \forall \beta\in(0,\gamma].$$

Suppose this is not so and that for every $\gamma>0$ there is a $\beta\in(0,\gamma]$ such that
$$\bar{y}+\beta n(\bar{y})\in\Sigma(C).$$

Then, for sufficiently small γ, $\bar{y}+\beta n(\bar{y})$ belongs to $N(\delta,\bar{y})$ so that
$$y=\bar{y}+\beta n(\bar{y})\in N(\delta,\bar{y})\cap\Sigma(C)$$

and, as a consequence of (11.1), we obtain
$$\bar{y}+\beta n(\bar{y})=\bar{y}+\Delta y+o(\|\Delta y\|),$$

where
$$\bar{y}+\Delta y\in T_{\Sigma(C)}(\bar{y}).$$

Consequently we have
$$\beta n(\bar{y})=\Delta y+o(\|\Delta y\|),$$

whence
$$\beta=\|\Delta y\|\left\|\frac{\Delta y}{\|\Delta y\|}+\frac{o(\|\Delta y\|)}{\|\Delta y\|}\right\| \qquad (11.3)$$

so that β and $\|\Delta y\|$ are of the same order of magnitude. But
$$n^T(\bar{y})\Delta y=0$$

so that $\beta=o(\|\Delta y\|)$. This is not possible since (11.3) implies that
$$\lim_{\beta\to 0}\frac{o(\|\Delta y\|)}{\beta}=0.$$

Thus there is a $\gamma>0$ such that, given $\beta\in(0,\gamma]$, $\bar{y}+\beta n(\bar{y})$ belongs *either* to $A/\Sigma(C)$ *or* to $B/\Sigma(C)$. Suppose that
$$\bar{y}+\beta_1 n(\bar{y})\in A/\Sigma(C), \qquad \beta_1\in(0,\gamma]$$

and
$$\bar{y} + \beta_2 n(\bar{y}) \in B/\Sigma(C), \quad \beta_2 \in (0, \gamma].$$

Then
$$\Phi[\bar{y} + \beta_1 n(\bar{y})] > C$$

and
$$\Phi[\bar{y} + \beta_2 n(\bar{y})] < C$$

so that, in view of the continuity of $\Phi(\cdot)$, there is a $\beta_3 \in (0, \gamma]$ such that
$$\Phi[\bar{y} + \beta_3 n(\bar{y})] = C$$

and hence that
$$\bar{y} + \beta_3 n(\bar{y}) \in \Sigma(C).$$

This is contrary to our earlier result that, given $\beta \in (0, \gamma]$, $\bar{y} + \beta n(\bar{y})$ belongs to $A/\Sigma(C)$ or $B/\Sigma(C)$. However, that still allows the possibility of *both* $n(\bar{y})$ and $-n(\bar{y})$ pointing into the *same* region. We show now that this cannot happen, thereby establishing (11.2).

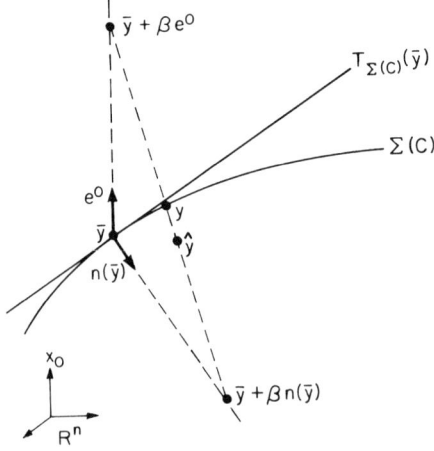

Figure 11.2. Construction in the proof of (11.4).

Chap. 11 • Regular Optimal Trajectories

Let e^0 denote the unit vector at $\bar{y}$ that points in the direction of increasing x_0. Then, if $n(\bar{y})$ points into $B/\Sigma(C)$ $[A/\Sigma(C)]$,

$$n_0(\bar{y}) \triangleq n^T(\bar{y})e^0 \leq 0 \quad [\geq 0]. \tag{11.4}$$

To prove (11.4) consider the points $\bar{y}+\beta e^0$ and $\bar{y}+\beta n(\bar{y})$, $\beta > 0$, and suppose that $n(\bar{y})$ points into $B/\Sigma(C)$; see Figure 11.2. Form the convex combination

$$\hat{y} = \alpha(\bar{y}+\beta e^0) + (1-\alpha)[\bar{y}+\beta n(\bar{y})], \quad \alpha \in [0,1]. \tag{11.5}$$

For sufficiently small β

$$\hat{y} \in N(\delta, \bar{y}) \quad \forall \alpha \in [0,1].$$

Recall the equation of $\Sigma(C)$; namely,

$$\Phi(y) = C.$$

Thus, in view of (10.26),

$$\Phi[\bar{y}+\beta n(\bar{y})] < C$$

and

$$\Phi[\bar{y}+\beta e^0] > C,$$

so that, as a consequence of the continuity of $\Phi(\cdot)$, there is an $\alpha \in (0,1)$, say $\alpha = \alpha(\beta)$, such that

$$y = \alpha(\beta)(\bar{y}+\beta e^0) + [1-\alpha(\beta)][\bar{y}+\beta n(\bar{y})] \in \Sigma(C).$$

Therefore using (11.1) results in

$$\alpha(\beta)(\bar{y}+\beta e^0) + [1-\alpha(\beta)][\bar{y}+\beta n(\bar{y})] = \bar{y} + \Delta y + o(\|\Delta y\|) \tag{11.6}$$

whence

$$k\beta = \left\| \frac{\Delta y}{\|\Delta y\|} + \frac{o(\|\Delta y\|)}{\|\Delta y\|} \right\| \|\Delta y\| \tag{11.7}$$

where $k \in (0, 1)$. In other words, β and $\|\Delta y\|$ are of the same order of magnitude. On multiplying (11.6) by $n^T(\bar{y})$ one obtains

$$\alpha(\beta)\left[n^T(\bar{y})e^0\right] = \alpha(\beta) - 1 + \frac{n^T(\bar{y})o(\|\Delta y\|)}{\beta}. \tag{11.8}$$

But $\alpha(\beta) \in (0, 1)$ and by (11.7)

$$\lim_{\beta \to 0} \frac{o(\|\Delta y\|)}{\beta} = 0$$

so that (11.8) implies (11.4). The same conclusion is reached if $n(\bar{y})$ points in $A/\Sigma(C)$ by employing $-e^0$ in place of e^0.

Now, provided $n_0(\bar{y}) \neq 0$, it follows from (11.4) that $-n(\bar{y})$ must point into $A/\Sigma(C)$ if $n(\bar{y})$ points into $B/\Sigma(C)$ so that one can always choose one of the two normals to point into $B/\Sigma(C)$, thereby establishing (11.2). If $n_0(\bar{y}) = 0$ we cannot employ (11.4) to reach this conclusion. If $n_0(y) = 0$ suppose that $n(\bar{y})$ as well as $-n(\bar{y})$ point into the same region, say into $B/\Sigma(C)$. Consider a point $\bar{y} + \beta_1 n(\bar{y})$, $\beta_1 > 0$. For sufficiently small β_1

$$\bar{y} + \beta_1 n(\bar{y}) \in B/\Sigma(C)$$

so that there is a $y^1 \in N(\delta, \bar{y}) \cap \Sigma(C)$ such that

$$y^1 = \bar{y} + \beta_1 n(\bar{y}) + \rho e^0, \quad \rho > 0.$$

Then, for sufficiently small β_1, it follows from the continuity of $\Phi(\cdot)$ that there is a $\beta_2 > 0$ such that

$$y^2 = \bar{y} - \beta_2 n(\bar{y}) + \rho e^0 \in N(\delta, \bar{y}) \cap \Sigma(C).$$

This is illustrated in Figure 11.3. But by (11.1) one has

$$y^1 = \bar{y} + \Delta y^1 + o(\|\Delta y^1\|),$$
$$y^2 = \bar{y} + \Delta y^2 + o(\|\Delta y^2\|),$$

with $\bar{y} + \Delta y^i$, $i = 1, 2$, in $T_{\Sigma(C)}(\bar{y})$. However,

$$\Delta y^1 = \Delta y^2 = \rho e^0 \stackrel{\triangle}{=} \Delta y$$

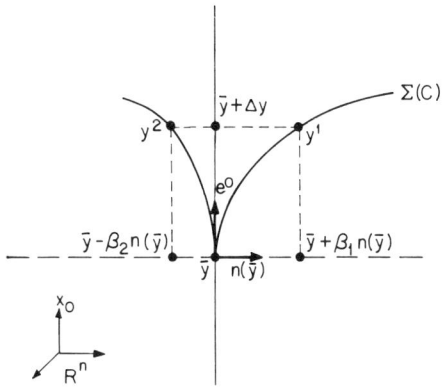

Figure 11.3. $n_0(\bar{y}) = 0$.

so that the one-to-one mapping property of (11.1) is contradicted. Thus not both $n(\bar{y})$ and $-n(\bar{y})$ can point into the same region. This concludes the proof of (11.2). Henceforth, we shall select the normal $n(\bar{y})$ that points into $B/\Sigma(C)$.

11.2. Necessary Conditions at a Regular Interior Point

We begin the discussion of necessary conditions for optimal control by considering the consequences of Theorem 10.1 at a regular interior point of a limiting surface.

Let $\Gamma^*(C)$ denote an optimal trajectory corresponding to the control $u^*(\cdot):[t_0, t_1^*] \to R^m$ that is optimal at x^0 and generates the solution

$$y^*(\cdot) = \begin{bmatrix} x_0^*(\cdot) \\ \hdashline x^*(\cdot) \end{bmatrix} : [t_0, t_1^*] \to R^{n+1}, \qquad y^*(t_0) = y^0,$$

of the augmented state equation (10.21). Suppose that the point $y^*(t)$ of $\Gamma^*(C)$ is a regular interior point of the limiting surface $\Sigma(C)$ containing $\Gamma^*(C)$. Let $n(y^*(t))$ be the unit vector that is normal to the tangent plane $T_{\Sigma(C)}(y^*(t))$ and points into $B/\Sigma(C)$. Then it is readily shown that

$$n^T(y^*(t)) h[y^*(t), u] \leq 0 \qquad \forall u \in U \tag{11.9}$$

and

$$n^T(y^*(t))h[y^*(t), u^*(t)] = 0. \tag{11.10}$$

If $u^*(\cdot)$ is discontinuous at $t \in (t_0, t_1^*)$ this condition holds for $u^*(t-0)$ and $u^*(t+0)$, the former if $n(y^*(\cdot))$ is continuous at t.

The relations (11.9) and (11.10) are "intuitively obvious" consequences of the basic properties embodied in Theorem 10.1. According to that theorem no trajectory emanating from a point $y(t) \in \Sigma(C)$ can have a point in $B/\Sigma(C)$. Thus the "velocity vector" $h[y(t), u(t)]$ must point into the region $A/\Sigma(C) \cup \Sigma(C)$ no matter what the control value $u(t) \in U$. The condition (11.9) is a statement of this consequence of Theorem 10.1 at a regular interior point of $\Sigma(C)$. Similarly, since $\Gamma^*(C)$ lies in $\Sigma(C)$ according to Theorem 10.1, the "velocity vector" $h[y^*(t), u^*(t)]$ must be tangent to $\Sigma(C)$ at a regular interior point $y^*(t)$; the relation (11.10) reflects this property.

The proof of (11.9) and (11.10) is analogous to that of (11.4). Let $\Gamma(\tilde{C})$ denote a trajectory corresponding to the admissible control $u(\cdot): [t, t_1] \to R^m$ and the solution $y(\cdot): [t, t_1] \to R^{n+1}$, $y(t) = y^*(t)$, of the augmented state equation (10.21); such a solution exists for $t_1 > t$. In other words, $\Gamma(\tilde{C})$ is a trajectory emanating from $y^*(t) \in \Sigma(C)$. Since $u(\cdot)$ is continuous at the initial point t of $[t, t_1]$, it follows from the definition of the derivative $\dot{y}(t)$, together with (10.21), that

$$y(t + \Delta t) = y(t) + h[y(t), u(t)] \Delta t + o(\Delta t) \tag{11.11}$$

where $\Delta t > 0$ and $\lim_{\Delta t \to 0} [o(\Delta t)/\Delta t] = 0$.

To establish (11.9) and (11.10) we consider the convex combination

$$\hat{y} = \alpha\{y^*(t) + h[y^*(t), u(t)]\Delta t + o(\Delta t)\} + (1-\alpha)[y^*(t) + n(y^*(t))\Delta t] \tag{11.12}$$

in place of (11.5); see Figure 11.4. The same arguments as those employed in the proof of (11.4) result in establishing (11.9) and (11.10). Finally, if $u^*(\cdot)$ is discontinuous at t we employ $u^*(t+0)$ to obtain (11.10). The validity of (11.10) for $u^*(t-0)$ follows from the continuity of $h(\cdot)$ and $x^*(\cdot)$, and the existence of $u^*(t-0)$, provided $n(y^*(\cdot))$ is continuous at t.

While the conditions (11.9) and (11.10) are *necessary* at a regular interior point—that is, must be satisfied—they are not as yet useful for deciding if a given control *may* be optimal, that is, is a candidate for an

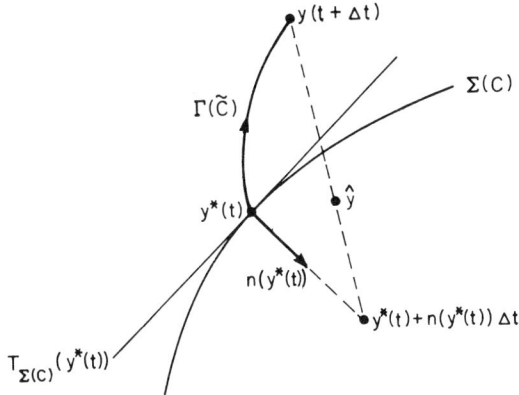

Figure 11.4. Construction in the proof of (11.9)–(11.10).

optimal control. These conditions involve the normal $n(y^*(t))$ of the tangent plane $T_{\Sigma(C)}(y^*(t))$. At this stage of discussion we have no information about $n(y^*(t))$. We turn now to an investigation of this normal, provided it is defined.

11.3. A Linear Transformation

In order to discuss the behavior of the normal vector $n(y^*(t))$ for $t \in [t_0, t_1^*]$ we shall employ a certain linear transformation. In this section we define this transformation and state some of its properties.

Consider the piecewise smooth function $\eta(\cdot):[t_0, t_1^*] \to R^{n+1}$ whose components $\eta_j(\cdot)$, $j = 0, 1, 2, \ldots, n$, are the solutions of

$$\dot{\eta}_j(t) = \sum_{i=0}^{n} \frac{\partial f_j[x^*(t), u^*(t)]}{\partial x_i} \eta_i(t) \qquad (11.13)$$

for given initial values $\eta_j(t_0) = \eta_j^0$. Equations (11.13) are the *variational equations* associated with the solution $y^*(\cdot)$ of the augmented state equation (10.21); for instance, see Ref. 11.1.

Equations (11.13) define a linear transformation $A(t, t_0)$ of $\eta^0 \triangleq [\eta_0^0 \ \eta_1^0 \ldots \eta_n^0]^T$ such that

$$\eta(t) = A(t, t_0) \eta^0, \qquad t \in [t_0, t_1^*]. \qquad (11.14)$$

Since equations (11.13) are linear and homogeneous, this transformation (matrix) is nonsingular; that is, it possesses an inverse so that

$$\eta^0 = A^{-1}(t, t_0)\eta(t)$$

for all $t \in [t_0, t_1^*]$; for instance, see Ref. 11.2. Recall that $u^*(\cdot)$ may have points of discontinuity where we may assign to $u^*(\cdot)$ its left or right limit; of course, $\eta(\cdot)$ is continuous at such points.

Next consider an n-dimensional plane $\Pi(y^*(t_0))$ containing the point $y^*(t_0)$. Such a plane is defined by the $(n+1)$-dimensional vectors from $y^*(t_0)$ to the points of the plane. Let $\varepsilon\eta^0$, $\varepsilon \in R^1$, be such a vector; see Figure 11.5. We are concerned with the transform $\Pi(y^*(t))$ of the plane $\Pi(y^*(t_0))$ by means of the linear transformation $A(t, t_0)$; that is,

$$\Pi(y^*(t)) \triangleq \{y \in R^{n+1} | y = y^*(t) + \varepsilon\eta(t), \, y^*(t_0) + \varepsilon\eta^0 \in \Pi(y^*(t_0))\}.$$

The following lemma embodies a property of interest to us.

Lemma 11.1. *The transform $\Pi(y^*(t))$ of the n-dimensional plane $\Pi(y^*(t_0))$ by means of $A(t, t_0)$ is defined for all $t \in [t_0, t_1]$; its dimension is the same as that of the plane $\Pi(y^*(t_0))$.*

Proof. Let $\Pi(y^*(t_0))$ be a plane of dimension n so that there is a basis of n linearly independent vectors η^{0k} of dimension $n+1$, $k = 1, 2, \ldots, n$; namely, given

$$y^*(t_0) + \varepsilon\eta^0 \in \Pi(y^*(t_0)), \qquad \varepsilon \in R^1,$$

there are n constants c_k, $k = 1, 2, \ldots, n$, such that

$$\eta^0 = \sum_{k=1}^{n} c_k \eta^{0k}.$$

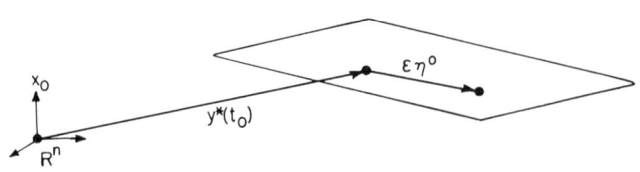

Figure 11.5. A plane at $y^*(t_0)$.

Chap. 11 • Regular Optimal Trajectories

Consequently the transform of η^0 by means of $A(t, t_0)$ is

$$\eta(t) = \sum_{k=1}^{n} c_k \eta^k(t),$$

where

$$\eta^k(t) = A(t, t_0) \eta^{0k}.$$

The vectors $\eta^k(t)$ are defined for all $t \in [t_0, t_1^*]$. Hence, to show that $\Pi(y^*(t))$ is defined and is a plane of dimension n, we need only prove that the vectors $\eta^k(t)$, $k = 1, 2, \ldots, n$, are linearly independent, that is, that there do not exist constants b_k, not all zero, such that

$$\sum_{k=1}^{n} b_k \eta^k(t) = 0.$$

Consider

$$\sum_{k=1}^{n} b_k \eta^k(t) = A(t, t_0) \sum_{k=1}^{n} b_k \eta^{0k}.$$

Since the inverse $A^{-1}(t, t_0)$ exists for all $t \in [t_0, t_1^*]$ we have

$$\sum_{k=1}^{n} b_k \eta^{0k} = A^{-1}(t, t_0) \sum_{k=1}^{n} b_k \eta^k(t). \tag{11.15}$$

However the vectors η^{0k}, $k = 1, 2, \ldots, n$, are linearly independent; that is, there do not exist constants b_k, not all zero, such that

$$\sum_{k=1}^{n} b_k \eta^{0k} = 0.$$

Thus, in view of (11.15), the vectors $\eta^k(t)$ are linearly independent. □

11.4. Transformation of the Tangent Plane

In preparation for deducing the behavior of the normal $n(y^*(t))$ along $\Gamma^*(C)$, we consider the transfer of the tangent plane $T_{\Sigma(C)}(y^*(t_0))$ by means of the linear transformation $A(t, t_0)$.

Let $y^*(t_0)$, the initial point of the optimal trajectory $\Gamma^*(C)$, be a regular interior point of the limiting surface $\Sigma(C)$, so that the tangent plane $T_{\Sigma(C)}(y^*(t_0))$ is defined. Then we have the following lemma.

Lemma 11.2. *If the points $y^*(t_0)$ and $y^*(t)$ of the optimal trajectory $\Gamma^*(C)$ are regular interior points of the limiting surface $\Sigma(C)$, then the tangent plane $T_{\Sigma(C)}(y^*(t_0))$ is transformed into the tangent plane $T_{\Sigma(C)}(y^*(t))$ by the linear transformation $A(t, t_0)$ defined by equation (11.13).*

Proof. Let $T(y^*(t))$ denote the transform of $T_{\Sigma(C)}(y^*(t_0))$ by $A(t, t_0)$. Since $y^*(t)$ is a regular interior point of $\Sigma(C)$, the tangent plane $T_{\Sigma(C)}(y^*(t))$ is defined. We wish to show that

$$T_{\Sigma(C)}(y^*(t)) = T(y^*(t)). \tag{11.16}$$

Consider a vector $\eta \stackrel{\Delta}{=} [\eta_0 \; \eta_1 \cdots \eta_n]^T$ such that

$$y^*(t) + \varepsilon\eta \in T(y^*(t)). \tag{11.17}$$

Then, since $A(t, t_0)$ is nonsingular, there is a vector $\eta^0 = A^{-1}(t, t_0)\eta$ where

$$y^*(t_0) + \varepsilon\eta^0 \in T_{\Sigma(C)}(y^*(t_0)). \tag{11.18}$$

Thus, for sufficiently small $\varepsilon > 0$, there is a point

$$y^*(t_0) + \varepsilon\eta^0 + o(\varepsilon) \in \Sigma(C). \tag{11.19}$$

Now consider the solution $y(\cdot):[t_0, t_1^*] \to R^{n+1}$, $y(t_0) = y^*(t_0) + \varepsilon\eta^0 + o(\varepsilon)$, generated by the control $u^*(\cdot):[t_0, t_1^*] \to R^m$. This solution exists for sufficiently small $\varepsilon > 0$ and is such that (see Ref. 11.3)

$$y(t) = y^*(t) + \varepsilon\eta(t) + o(t, \varepsilon), \tag{11.20}$$

where

$$\eta(t) = A(t, t_0)\eta^0 \tag{11.21}$$

and $[o(t, \varepsilon)/\varepsilon] \to 0$ uniformly for all $t \in [t_0, t_1^*]$ as $\varepsilon \to 0$. In view of (11.17), (11.18) and (11.21), we have

$$\eta(t) = \eta. \tag{11.22}$$

Chap. 11 • Regular Optimal Trajectories

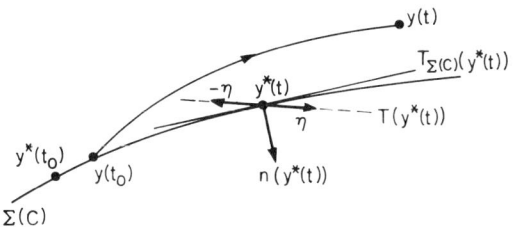

Figure 11.6. Construction in the proof of Lemma 11.2.

Let $n(y^*(t))$ denote the unit vector that is normal to $T_{\Sigma(C)}(y^*(t))$ and points into $B/\Sigma(C)$. Recall now that, as a consequence of Theorem 10.1,

$$y(t) \in A/\Sigma(C) \cup \Sigma(C). \tag{11.23}$$

This is illustrated in Figure 11.6. Employing an argument that is entirely analogous to the one used to prove condition (11.9), one can show that

$$n^T(y^*(t))\eta \leq 0. \tag{11.24}$$

Upon repeating the procedure leading to (11.24) with $-\eta$ in place of η, one arrives at

$$n^T(y^*(t))(-\eta) \leq 0. \tag{11.25}$$

It follows that

$$n^T(y^*(t))\eta = 0, \tag{11.26}$$

and hence that

$$y^*(t) + \varepsilon\eta \in T_{\Sigma(C)}(y^*(t)). \tag{11.27}$$

But η is an arbitrary vector in $T(y^*(t))$, and so (11.16) is established. □

11.5. Regular Optimal Trajectories

We turn now to a discussion of necessary conditions (11.9) and (11.10) for *regular* optimal trajectories, that is, for optimal trajectories all of whose points, with the possible exception of the terminal point, are regular interior points of a limiting surface.

Suppose that $\Gamma^*(C)$ is a regular optimal trajectory corresponding to the solution $y^*(\cdot): [t_0, t_1^*] \to R^{n+1}$, so that $y^*(t)$ is a regular interior point of $\Sigma(C)$ for all $t \in [t_0, t_1^*)$. Then, if η^0 is a vector in $T_{\Sigma(C)}(y^*(t_0))$, its transform $\eta(t) = A(t, t_0)\eta^0$ is a vector in $T_{\Sigma(C)}(y^*(t))$ according to Lemma 11.2.

Now consider the piecewise smooth function $\lambda(\cdot): [t_0, t_1^*] \to R^{n+1}$ whose components $\lambda_j(\cdot)$, $j = 0, 1, \ldots, n$, are the solutions of the so-called *adjoint equations*

$$\dot{\lambda}_j(t) = -\sum_{i=0}^{n} \frac{\partial f_i[x^*(t), u^*(t)]}{\partial x_j} \lambda_i(t) \tag{11.28}$$

for particular initial conditions $\lambda_j(t_0) = \lambda_j^0$. The function $\lambda(\cdot)$ is called the *adjoint*, or sometimes the *costate*. Note that $\lambda(t) \neq 0$ for all $t \in [t_0, t_1^*]$ if $\lambda(\bar{t}) \neq 0$, $\bar{t} \in [t_0, t_1^*]$, since the equations (11.28) are linear and homogeneous and hence define a nonsingular linear transformation.

Equations (11.28) are adjoint to the variational equations (11.13) since, given any solution $\eta(\cdot)$ of (11.13) and any solution $\lambda(\cdot)$ of (11.28), we have

$$\frac{d}{dt}[\lambda^T(t)\eta(t)] = 0$$

so that

$$\lambda^T(t)\eta(t) = \text{constant} \qquad \forall t \in [t_0, t_1^*]. \tag{11.29}$$

Then, if η^0 belongs to $T_{\Sigma(C)}(y^*(t_0))$ and λ^0 is nonzero and in the direction of the normal $n(y^*(t_0))$, the vectors η^0 and λ^0 are orthogonal and

$$\lambda^{0T}\eta^0 = 0$$

so that (11.29) implies

$$\lambda^T(t)\eta(t) = 0 \qquad \forall t \in [t_0, t_1^*]. \tag{11.30}$$

Now, since η^0 may be any one of n linearly independent vectors in $T_{\Sigma(C)}(y^*(t_0))$ and since the transforms, $\eta(t)$, of these η^0 remain linearly independent and in the tangent plane $T_{\Sigma(C)}(y^*(t))$ for $t \in [t_0, t_1^*)$, it follows from (11.30) that $\lambda(t)$ is normal to $T_{\Sigma(C)}(y^*(t))$. Hence the adjoint equations (11.28) define the evolution of a nonzero vector $\lambda(t)$ that is normal to $T_{\Sigma(C)}(y^*(t))$, where $\lambda(t_0)$ is nonzero and normal to $T_{\Sigma(C)}(y^*(t_0))$. Having chosen $\lambda(t_0)$ in the direction of the normal $n(y^*(t_0))$, that is, pointing into

$B/\Sigma(C)$, it is readily shown that $\lambda(t)$ points into $B/\Sigma(C)$ for all $t \in [t_0, t_1^*)$. Since $\partial f_i(x, u)/\partial x_0 \equiv 0$ it follows that $\dot{\lambda}_0(t) \equiv 0$ so that

$$\lambda_0(t) = \text{constant} \quad \forall t \in [t_0, t_1^*].$$

Thus, if $\lambda_0(t_0) < 0$ then $\lambda_0(t) < 0$ for all $t \in [t_0, t_1^*)$ so that (11.4) implies that $\lambda(t)$ points into $B/\Sigma(C)$. If $\lambda_0(t_0) = 0$ so that $\lambda_0(t) \equiv 0$, the conclusion follows from the continuity and nonvanishing of $\lambda(\cdot)$ together with a result in Section 11.1; namely, $\lambda(t)$ points into $B/\Sigma(C)$ or into $A/\Sigma(C)$ but not into both regions.

We conclude that there exists a nonzero solution $\lambda(\cdot)$ of (11.28) such that $\lambda(t)$ is codirectional with $n(y^*(t))$. Consequently the conditions (11.9) and (11.10) must be satisfied for such a normal vector; that is, for all $t \in [t_0, t_1^*)$

$$\lambda^T(t) h[y^*(t), u] \leq 0 \quad \forall u \in U \quad (11.31)$$

and

$$\lambda^T(t) h[y^*(t), u^*(t)] = 0, \quad (11.32)$$

where (11.32) applies with $u^*(t-0)$ and $u^*(t+0)$ if $u^*(\cdot)$ is discontinuous at $t \in (t_0, t_1^*)$.

Furthermore, as noted before,

$$\lambda_0(t) = \text{constant} \leq 0 \quad \forall t \in [t_0, t_1^*]. \quad (11.33)$$

Finally, since $h(\cdot)$ is continuous on $R^{n+1} \times R^m$, $\lambda(\cdot)$ and $y^*(\cdot)$ are continuous on $[t_0, t_1^*]$, and $u^*(\cdot)$ is continuous at t_1^*, the relations (11.31), (11.32) apply for $t \in [t_0, t_1^*]$.

While the conditions (11.31) and (11.32) constitute an improvement over (11.9) and (11.10) in that the adjoint equations (11.28) define the evolution of $\lambda(t)$ for a given λ^0, we do not have as yet a method for choosing a particular solution for which these conditions must be satisfied.

The variables $x_0(t)$ and $\lambda_0(t)$ are not of concern because the former does not occur in (11.31)–(11.32) nor in the state equations (10.1) and adjoint equations (11.28), and the latter is known to be a nonpositive constant; that is, either $\lambda_0(t) \equiv 0$ or it is a negative constant. In the latter case one may divide (11.31)–(11.32) by $-\lambda_0(t)$ without affecting their validity. This amounts merely to rescaling $\lambda(t)$. Thus one need consider only two possibilities, $\lambda_0(t) \equiv 0$ and $\lambda_0(t) \equiv -1$ (or any other negative constant).

To verify the conditions (11.31)–(11.32) for a given control $u^*(\cdot)$ one must integrate a set of $2n$ ordinary (and usually coupled) first-order differential equations, namely, (10.1) and (11.28) for $j=1,2,\ldots,n$. Thus, for a well-posed problem, we require $2n$ independent end conditions. Recall the imposed end conditions (10.3): A solution $x^*(\cdot): [t_0, t_1^*] \to R^n$ generated by $u^*(\cdot)$ must be such that

$$\begin{aligned} x^*(t_0) &= x^0 \quad \text{(given)}, \\ x^*(t_1^*) &\in \theta^1 \quad \text{(given)}. \end{aligned}$$

If the target set θ^1 is a point, say x^1, then there are indeed $2n$ independent end conditions, n initial and n terminal ones. If θ^1 is not a single point in the state space R^n, nothing more can be said without additional information about the target set θ^1. Before turning to a discussion of a certain class of target sets, let us pause briefly to illustrate the ideas introduced in the preceding sections.

11.6. An Illustrative Example

Recall the problem treated in Section 10.8. Note first of all that every point of an optimal trajectory $\Gamma^*(C)$, with the exception of the terminal point

$$\begin{bmatrix} x_0^*(t_1^*) \\ x_1^*(t_1^*) \end{bmatrix} = \begin{bmatrix} C \\ 0 \end{bmatrix}$$

is a regular interior point of $\Sigma(C)$; thus, $\Gamma^*(C)$ is a regular optimal trajectory.

The conditions (11.9)–(11.10) are verified by noting that

$$n(y^*(t)) = \frac{\sqrt{2}}{2} \begin{bmatrix} -1 \\ -\operatorname{sgn} x_1^0 \end{bmatrix}$$

while

$$h[y^*(t), u] = \begin{bmatrix} 1 \\ u \end{bmatrix},$$

$$h[y^*(t), u^*(t)] = \begin{bmatrix} 1 \\ -\operatorname{sgn} x_1^0 \end{bmatrix}.$$

Lemma 11.2 is satisfied trivially since $y^*(t) + \varepsilon\eta(t) + o(t, \varepsilon) \in \Sigma(C)$ at every regular interior point of $\Sigma(C)$.

The variational equations (11.13) are

$$\dot{\eta}_0(t) = 0, \quad \dot{\eta}_1(t) = 0$$

so that $A(t, t_0)$ is the identity transformation; that is,

$$\eta(t) = \eta^0 \quad \forall t \in [t_0, t_1^*].$$

It follows that

$$y^*(t_0) + \epsilon\eta^0 \in T_{\Sigma(C)}(y^*(t_0)) \Rightarrow y^*(t) + \epsilon\eta(t) \in T_{\Sigma(C)}(y^*(t))$$

since $T_{\Sigma(C)}(y^*(t))$ is merely the line having direction η^0.

The adjoint equations (11.28) are

$$\dot{\lambda}_0(t) = 0, \quad \dot{\lambda}_1(t) = 0$$

so that

$$\lambda_0(t) \equiv \text{constant}, \quad \lambda_1(t) \equiv \text{constant}.$$

Hence there is a nonzero $\lambda(t)$ for which (11.31)–(11.32) are met; for instance,

$$\lambda_0(t) \equiv -1, \quad \lambda_1(t) \equiv -\text{sgn } x_1^0.$$

11.7. The Terminal Transversality Condition

In this section we derive conditions which must be satisfied at the terminal point of an optimal trajectory for a certain class of target sets.

Suppose that the target set θ^1 is a nonempty smooth $(n-q)$-dimensional manifold; that is, it is the intersection of $q \leq n$ smooth surfaces

$$\theta^1 \triangleq \{x \in R^n | \theta_i^1(x) = 0, i = 1, 2, \ldots, q\}, \tag{11.34}$$

where the $\theta_i^1(\cdot) : R^n \to R^1$ are prescribed functions of class C^1. We assume also that the matrix

$$\begin{bmatrix} \dfrac{\partial \theta_1^1(x)}{\partial x_1} & \cdots & \dfrac{\partial \theta_1^1(x)}{\partial x_n} \\ \vdots & & \vdots \\ \dfrac{\partial \theta_q^1(x)}{\partial x_1} & \cdots & \dfrac{\partial \theta_q^1(x)}{\partial x_n} \end{bmatrix}$$

has maximum rank, namely, q, at every $x \in \theta^1$, or equivalently that the vectors grad $\theta_i^1(x)$, $i = 1, 2, \ldots, q$, are linearly independent. This condition assures that the tangent plane of θ^1 exists at every $x \in \theta^1$; it is

$$T_{\theta^1}(x) = T_{\theta_1^1}(x) \cap T_{\theta_2^1}(x) \cap \cdots \cap T_{\theta_q^1}(x)$$

where $T_{\theta_i^1}(x)$ is the tangent plane of the smooth surface θ_i^1 with the normal vector grad $\theta_i^1(x)$.

Next consider the set

$$\Theta(C) \triangleq \{y \in R^{n+1} | x_0 = C, x \in \theta^1\} \tag{11.35}$$

so that

$$\Theta(C) \subset \Theta^1.$$

The tangent plane of $\Theta(C)$ at a point

$$\bar{y} = \begin{bmatrix} C \\ \bar{x} \end{bmatrix}$$

is

$$T_{\Theta(C)}(\bar{y}) = \{y \in R^{n+1} | x_0 = C, x \in T_{\theta^1}(\bar{x})\}.$$

Consider also the set

$$\Sigma(C) \cap \Theta^1 \triangleq \{y \in \mathcal{E}^* \cup \Theta^* | x_0 + V^*(x) = C, x \in \theta^1\}. \tag{11.36}$$

If $y^*(t_1^*)$ is the terminal point of the optimal trajectory $\Gamma^*(C)$ then

$$y^*(t_1^*) \in \Sigma(C) \cap \Theta^1$$

and

$$y^*(t_1^*) \in \Theta(C),$$

since

$$V^*(x) = 0 \quad \forall x \in \theta^* \subset \theta^1.$$

Furthermore recall the discussion at the end of Section 10.2, which implies that

$$V^*(x) \leq 0 \quad \forall x \in \theta^1 \cap [E^* \cup \theta^*].$$

Chap. 11 • Regular Optimal Trajectories

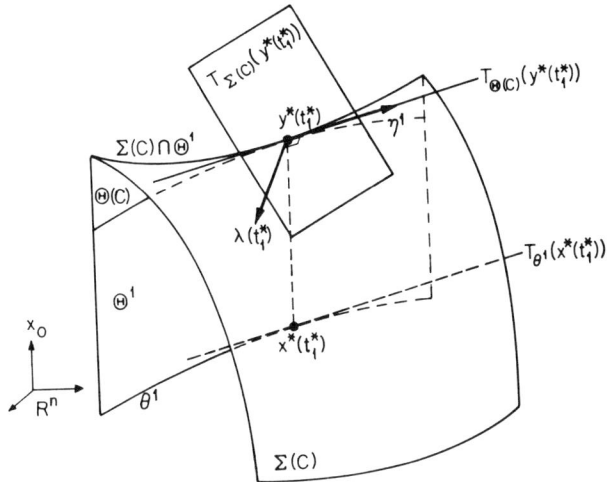

Figure 11.7. The terminal transversality condition.

Then it follows from (11.35) and (11.36) that

$$y \in \Theta(C) \Rightarrow y \notin A/\Sigma(C). \tag{11.37}$$

This is illustrated in Figure 11.7.

Now suppose that $y^*(t_1^*)$ is a regular interior point of $\Sigma(C)$. Then one can employ arguments which are similar to those used to establish (11.9)–(11.10) in order to show that (11.37) implies that

$$T_{\Theta(C)}(y^*(t_1^*)) \subset T_{\Sigma(C)}(y^*(t_1^*)).$$

Furthermore, since $y^*(t_1^*)$ is assumed to be a regular interior point of $\Sigma(C)$, $\lambda(t_1^*)$ is codirectional with the normal $n(y^*(t_1^*))$ of $T_{\Sigma(C)}(y^*(t_1^*))$; consequently, $\lambda(t_1^*)$ is normal to every vector $\eta^1 \triangleq [\eta_0^1 \ \eta_1^1 \cdots \eta_n^1]^T$ such that

$$y^*(t_1^*) + \varepsilon \eta^1 \in T_{\Theta(C)}(y^*(t_1^*)).$$

In other words, we arrive at

$$\lambda^T(t_1^*)\eta^1 = 0 \tag{11.38}$$

for all η^1 such that

$$\eta_0^1 = 0,$$

$$\sum_{i=1}^{n} \frac{\partial \theta_j^1(x^*(t_1^*))}{\partial x_i} \eta_i^1 = 0, \quad j = 1, 2, \ldots, q. \tag{11.39}$$

This is the terminal transversality condition; it is illustrated in Figure 11.7. As a result of the linear independence of the grad $\theta_i^1(x)$, $i = 1, 2, \ldots, q$, the equations (11.39) can be solved for $q+1$ components of η^1 as linear combinations of the remaining $n-q$ ones which are *arbitrary*. Upon substituting for these $q+1$ components in (11.38) one is left with $n-q$ terms, each of which is multiplied by an arbitrary component of η^1; the coefficients of these arbitrary components must be zero. This yields $n-q$ relations involving $x^*(t_1^*)$ and $\lambda(t_1^*)$; in fact, $\lambda_0(t_1^*)$ does not appear in these conditions. Thus the end conditions (10.3), together with the $n-q$ relations arising from the terminal transversality condition, provide $2n$ end conditions for $x_i(t)$ and $\lambda_i(t)$, $i = 1, 2, \ldots, n$.

If the terminal state is prescribed, $\theta^1 = \{x^1\}$, so that

$$\theta_i^1(x) = x_i - x_i^1, \quad i = 1, 2, \ldots, q = n,$$

then the terminal transversality condition is trivially satisfied.

11.8. A Maximum Principle

Before summarizing the necessary conditions for optimal control obtained thus far, we define the function $H(\cdot): R^{n+1} \times R^{n+1} \times R^m \to R^1$ by

$$H(\lambda, y, u) \triangleq \lambda^T h(y, u). \tag{11.40}$$

We are ready now to state a theorem.

Theorem 11.1. *If the control* $u^*(\cdot): [t_0, t_1^*] \to R^m$ *is optimal at the initial state* x^0 *and generates the regular optimal trajectory* $\Gamma^*(C)$ *corresponding to the solution* $y^*(\cdot): [t_0, t_1^*] \to R^{n+1}$ *of the augmented state equation (10.21), then there exists a nonzero solution* $\lambda(\cdot): [t_0, t_1^*] \to R^{n+1}$ *of the adjoint equations (11.28) such that*

(a) $\max_{u \in U} H[\lambda(t), y^*(t), u] = H[\lambda(t), y^*(t), u^*(t)]$,

(b) $H[\lambda(t), y^*(t), u^*(t)] = 0$,

(c) $\lambda_0(t) = \text{constant} \leq 0$

for all $t \in [t_0, t_1^*]$ where $u^*(\cdot)$ is continuous. If $u^*(\cdot)$ is discontinuous at $\bar{t} \in (t_0, t_1^*)$, (a) and (b) hold with $u^*(\bar{t}-0)$ and $u^*(\bar{t}+0)$.

(d) *Furthermore, if the terminal point $y^*(t_1^*)$ is a regular interior point of the limiting surface $\Sigma(C)$, and if the target set θ^1 is a smooth manifold, then the terminal transversality condition (11.38)–(11.39) is satisfied.*

Proof. The theorem is merely a summary of the results in Sections 11.5 and 11.7. □

11.9. Remarks

The following remarks concerning Theorem 11.1 are worth noting.

(i) For the solution $x^*(\cdot): [t_0, t_1^*] \to R^n$, $x^*(t_0) = x^0$, of the state equation (10.2), generated by the optimal control $u^*(\cdot)$, the state and adjoint equations can be written in the form

$$\dot{x}_j^*(t) = \frac{\partial H[\lambda(t), y^*(t), u^*(t)]}{\partial \lambda_j},$$

$$\dot{\lambda}_j(t) = -\frac{\partial H[\lambda(t), y^*(t), u^*(t)]}{\partial x_j}$$

(11.41)

for

$$j = 1, 2, \ldots, n, \qquad t \in [t_0, t_1^*].$$

As noted earlier, we are not concerned with $x_0^*(t)$ and $\lambda_0(t)$. Of course, as expected,

$$\dot{x}_0^*(t) = \frac{\partial H[\lambda(t), y^*(t), u^*(t)]}{\partial \lambda_0} = f_0[x^*(t), u^*(t)],$$

$$\dot{\lambda}_0(t) = -\frac{\partial H[\lambda(t), y^*(t), u^*(t)]}{\partial x_0} = 0.$$

(ii) As deduced here, Theorem 11.1 is a statement of *necessary* conditions for optimal control provided the corresponding trajectory is *regular*. It can be shown that the theorem is valid even in the absence of regularity;

this can be done by a generalization of the geometric approach (for instance, see Ref. 11.4) or by other methods (for instance, see Refs. 11.5 and 11.6).

(iii) As stated, Theorem 11.1 is a *maximum* principle in the sense that the value of $H[\lambda(t), y^*(t), \cdot]$ must be *maximized* by the value of an optimal control that *minimizes* the cost. This convention is kept here for historical reasons (see Ref. 11.5). Some authors state the necessary conditions as a *minimum* principle. This is readily accomplished by considering the normal $n(y^*(t))$ that points into the region $A/\Sigma(C)$ rather than $B/\Sigma(C)$ as we have done; in that case, $\lambda_0(t) = \text{constant} \geq 0$.

(iv) Before turning to a discussion of the utilization of Theorem 11.1, we comment briefly on its relation to dynamic programming; for instance, see Refs. 11.7–11.9.

Again let $\Phi(\cdot): \mathcal{E}^* \cup \Theta^* \to R^1$ be given by

$$\Phi(y) \triangleq x_0 + V^*(x)$$

so that, in accordance with the definition (10.24), a limiting surface is

$$\Sigma(C) \triangleq \{ y \in \mathcal{E}^* \cup \Theta^* | \Phi(y) = C \}.$$

Now consider $\bar{y} \in \Sigma(C)$ and suppose that $\Phi(\cdot)$ is continuously differentiable in a neighborhood of $\bar{y}$; that is, $\Phi(\cdot)$ is of class C^1 on $N(\delta, \bar{y})$ for some $\delta > 0$. Then the tangent plane $T_{\Sigma(C)}(\bar{y})$ is defined. The converse need not be so; that is,

$$\text{grad } \Phi(\bar{y}) \triangleq \left[1 \quad \frac{\partial V^*(\bar{x})}{\partial x_1} \quad \cdots \quad \frac{\partial V^*(\bar{x})}{\partial x_n} \right]^T$$

need not be defined even though $T_{\Sigma(C)}(\bar{y})$ exists (as in the case of $T_{\Sigma(C)}(\bar{y})$ being x_0-cylindrical).

Let $\bar{y} = y^*(t)$ be a point of $\Gamma^*(C)$; then $n(y^*(t))$ is codirectional with $\lambda(t)$ and $\lambda_0(t) < 0$ since $n_0(y^*(t)) \neq 0$ in view of our supposition about the existence of grad $\Phi[y^*(t)]$. It follows that

$$\lambda(t) = \lambda_0(t) \text{ grad } \Phi[y^*(t)] \tag{11.42}$$

since grad $\Phi(y)$ is directed opposite to the normal $n(y)$. If we choose

$\lambda_0(t) = -1$, as we may, then

$$\lambda_j(t) = -\frac{\partial V^*[x^*(t)]}{\partial x_j}, \quad j = 1, 2, \ldots, n. \tag{11.43}$$

By employing (11.43) in the conditions (a) and (b) of Theorem 11.1, and noting that every point of $\mathcal{E}^* \cup \Theta^*$ belongs to an optimal trajectory, we arrive at the fundamental equation of *dynamic programming*

$$\min_{u \in U} \left[f_0(x, u) + \sum_{j=1}^{n} \frac{\partial V^*(x)}{\partial x_j} f_j(x, u) \right] = 0 \tag{11.44}$$

for $x \in \mathcal{E}^* \cup \Theta^*$. This partial differential equation for $V^*(\cdot)$ is subject to the boundary condition $V^*(x) = 0$ for $x \in \theta^*$. For a further discussion of this equation the reader should consult Refs. 11.7–11.9 and Section 16.2.

11.10. Extremal Control

A control that is feasible at the initial state x^0 and for which the conditions of Theorem 11.1 are satisfied is termed *extremal* at x^0. We must keep in mind that Theorem 11.1 is a statement of *necessary* conditions for optimality. Thus an extremal control need not be an optimal one. All one can assert is that an extremal control *may* be optimal; it is only a candidate. Of course more than one extremal control can be optimal; that is, optimal control need not be unique.

To determine if an extremal control is indeed optimal one can proceed in one of two ways. The first of these utilizes *existence theorems*; namely, one invokes a theorem, the satisfaction of which assures that an optimal control exists (for instance, see Ref. 11.10). Next one deduces *all* extremal controls, that is, all feasible controls which *may* be optimal. Then an optimal control is an extremal one that yields the lowest cost in the set of costs due to *all* extremal controls. This approach has a number of drawbacks, among them being the requirement of finding *all* extremal controls. The second method involves the employment of *sufficient conditions*, that is, conditions whose satisfaction by a control assures that it is optimal. The main drawback of this technique is the restrictiveness of the imposed conditions, the satisfaction of which is required to guarantee optimality. Neither method can be used to determine that a particular control is *not* optimal. Hence a control may be optimal and yet fail to meet the conditions imposed by either existence or sufficiency theorems. Since existence theorems lie outside the scope of our treatment, we shall employ the second method in our subsequent discussion; see Chapter 15.

Since both methods for determining the optimality of a particular control require candidates, that is, controls for which the conditions of a sufficiency theorem can be tested or which are members of the set of all possibly optimal controls, one is interested in the *constructive* utilization of Theorem 11.1. In the following chapter, we present some examples of the use of Theorem 11.1 to deduce extremal controls. However, before doing so we revisit the simple illustrative example of Section 10.8.

11.11. An Illustrative Example

Recall once more the example of Section 10.8. There we surmised that the control

$$u^*(t) \equiv -\operatorname{sgn} x_1^0, \qquad t \in \left[0, |x_1^0|\right], \qquad (11.45)$$

is optimal. Let us verify the conditions of Theorem 11.1. Here the $H(\cdot)$ function is given by

$$H(\lambda, y, u) = \lambda_0 + \lambda_1 u$$

and therefore the adjoint equations are

$$\dot{\lambda}_0(t) = 0, \qquad \dot{\lambda}_1(t) = 0$$

so that

$$\lambda_0(t) \equiv \text{constant}, \qquad \lambda_1(t) \equiv \text{constant}.$$

Moreover the condition (a) of the theorem implies that

$$u^*(t) \equiv -1 \qquad \text{for } \lambda_1(t) < 0,$$
$$u^*(t) \equiv 1 \qquad \text{for } \lambda_1(t) > 0,$$

and by (b) neither $\lambda_0(t)$ nor $\lambda_1(t)$ can be zero, lest both are zero, which violates the condition requiring a nonzero solution $\lambda(\cdot)$. Thus we have

$$\lambda_1(t) = -|\lambda_1(t)| \operatorname{sgn} x_1^0.$$

Of course, the condition (c) is also satisfied.

Chap. 11 • Regular Optimal Trajectories

We have shown that $u^*(\cdot)$ is indeed extremal. By the same token, had we not "known" that $u^*(\cdot)$ is optimal, we could have employed Theorem 11.1 to deduce an extremal control given by (11.45) since that control is one—indeed, the only one—for which the conditions of the theorem are met.

Exercises

11.1. Recall the proof of Lemma 11.2. Show that (11.23) implies (11.24).

11.2. Utilizing equations (11.13) and (11.28), show that

$$\frac{d}{dt}\left[\lambda^T(t)\eta(t)\right]=0.$$

11.3. Recall Exercise 10.2. For that problem
   (a) show that optimal trajectories are regular;
   (b) verify the conditions (11.9)–(11.10);
   (c) discuss the behavior of the solution $\eta(\cdot)$ of (11.13);
   (d) verify the existence of an adjoint vector $\lambda(t)$ with which (11.31)–(11.32) are satisfied.

11.4. For the problem of Exercise 10.2 deduce an extremal control by means of the maximum principle, Theorem 11.1.

12

Examples of Extremal Control

12.1. Time-Optimality for a Constant-Power Rocket

It can be shown (for instance, see Ref. 12.1) that the equations of rectilinear motion of a rocket operating at constant propulsive power are

$$\dot{x}_1(t) = u(t), \qquad \dot{x}_2(t) = u^2(t) \tag{12.1}$$

where t is time, x_1 is velocity, and x_2 is inversely proportional to the mass of the rocket, while u is the acceleration caused by the thrust. This system is a model of a power-limited rocket, such as one powered by an ion engine, moving in force-free space. Since, in addition to the power limitation, the thrust acceleration is bounded in magnitude, we impose the (normalized) control constraint set

$$U = \{ u \in R^1 \mid |u| \leq 1 \}. \tag{12.2}$$

The problem is: Given the initial velocity and mass, that is, x_1^0 and x_2^0, it is desired to minimize the time, $t_1 - t_0$, required to reach a given velocity, x_1^1, while expending a given amount of mass, that is, changing $x_2(t)$ to a given value, x_2^1. Thus the cost is

$$\int_{t_0}^{t_1} dt = t_1 - t_0. \tag{12.3}$$

We utilize Theorem 11.1 in order to find extremal control(s).
We require the function $H(\cdot)$; in view of (12.1) and (12.3) it is given by

$$H(\lambda, y, u) = \lambda_0 + \lambda_1 u + \lambda_2 u^2 \tag{12.4}$$

Table 12.1. Extremal Controls

Case	Condition	Extremal control
(i)	$\lambda_1 > 0,\ \lambda_2 \geq 0$	$u(t) \equiv 1$
(ii)	$\lambda_1 < 0,\ \lambda_2 \geq 0$	$u(t) \equiv -1$
(iii)	$\lambda_1 < 0,\ \lambda_2 < 0$	$u(t) \equiv -\lambda_1/2\lambda_2$ or -1
(iv)	$\lambda_1 > 0,\ \lambda_2 < 0$	$u(t) \equiv -\lambda_1/2\lambda_2$ or 1
(v)	$\lambda_1 = 0,\ \lambda_2 < 0$	$u(t) \equiv 0$
(vi)	$\lambda_1 = 0,\ \lambda_2 > 0$	$|u(t)| \equiv 1$

so that the adjoint equations for $j = 1, 2$ are

$$\dot{\lambda}_1(t) = 0, \qquad \dot{\lambda}_2(t) = 0,$$

whence

$$\lambda_1(t) \equiv \text{constant} \stackrel{\triangle}{=} \lambda_1, \qquad \lambda_2(t) \equiv \text{constant} \stackrel{\triangle}{=} \lambda_2. \qquad (12.5)$$

Also, by (c) of Theorem 11.1, we have

$$\lambda_0(t) \equiv \text{constant} \stackrel{\triangle}{=} \lambda_0 \leq 0. \qquad (12.6)$$

As a consequence of (a) of the theorem, an *extremal* control $u(\cdot)$, generating the solution $y(\cdot)$, is a feasible control whose value $u(t)$ yields the

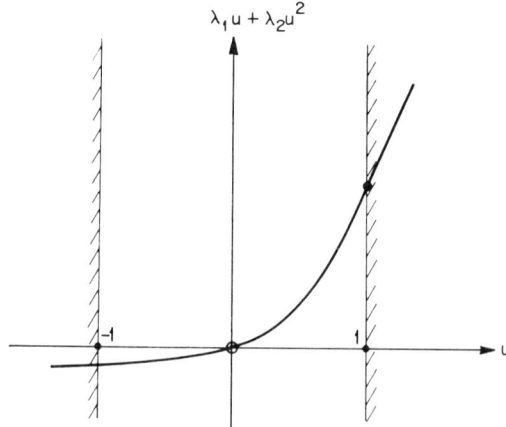

Figure 12.1. Case (i).

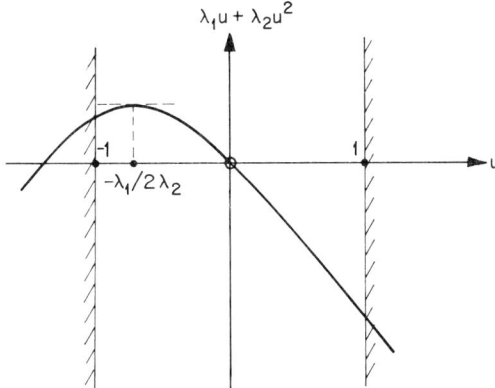

Figure 12.2. Case (iii) for $|\lambda_1/2\lambda_2|<1$.

maximum of

$$H[\lambda(t), y(t), u] = \lambda_0 + \lambda_1 u + \lambda_2 u^2 \qquad (12.7)$$

over U. The possible cases are listed in Table 12.1. The *possibly* extremal controls listed in this table are a consequence of the requirement of (a) of the theorem. To illustrate this consider, for example, cases (i), (iii), and (vi). Cases (ii) and (iv) are deduced similarly, while (v) can be ruled out except in the trivial situation of $x_i^1 = x_i^0$ for which $t_1 - t_0 = 0$. The relevant portion of $H[\lambda(t), y(t), u]$ as a function of u is shown in Figures 12.1–12.3. In cases (iii) or (iv), $u(t) \equiv -1$ or $u(t) \equiv 1$ if $|\lambda_1/2\lambda_2| \geq 1$.

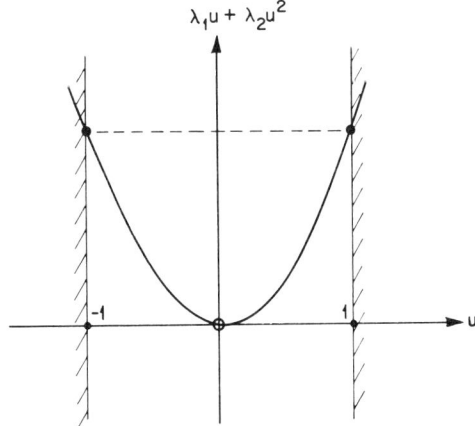

Figure 12.3. Case (vi).

While it appears as if there were six cases to be considered, in fact there are only two:

$$u(t) \equiv \text{constant}, \quad \text{(i)-(v)},$$
$$|u(t)| \equiv 1, \quad \text{(vi)}.$$

For cases (i)–(v) the constant value of $u(t)$ must satisfy constraint (12.2) and must be such that $u(\cdot)$ is feasible at x^0.

Case (vi) is an example of so-called *bang–bang* control since its values can *switch* from 1 to -1, or from -1 to 1, at a finite number of instances during the transfer.

Except in the trivial case (v), it follows from (12.1) with (12.2) that

$$\left| \frac{dx_2(t)}{dx_1(t)} \right| \leq 1 \tag{12.8}$$

so that, in view of the second of (12.1), the solutions are subject to the slope restriction (see Figure 12.4)

$$0 < \frac{dx_2(t)}{dx_1(t)} \leq 1 \quad \text{or} \quad -1 \leq \frac{dx_2(t)}{dx_1(t)} < 0. \tag{12.9}$$

As a consequence of the slope restriction, the region of initial states, $x \neq x^1$, from which an admissible control can transfer the state to the given terminal one is the open half-plane given by $x_2 < x_2^1$; in other words we have

$$E = \{ x \in R^2 \mid x_2 < x_2^1 \} \cup \{ x^1 \}. \tag{12.10}$$

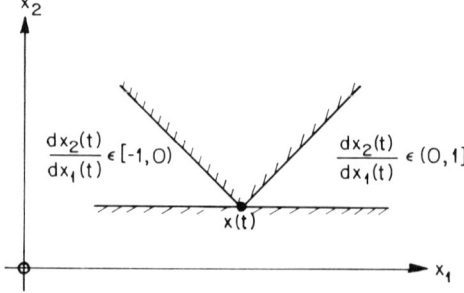

Figure 12.4. Slope constraint.

Furthermore it follows from (12.9) that E is divided into three subregions (see Figure 12.5):

$$I \stackrel{\triangle}{=} \{x \in R^2 | x_2 - x_2^1 \leq -|x_1 - x_1^1|\},$$

$$II \stackrel{\triangle}{=} \{x \in R^2 | x_1 < x_1^1, x_1 - x_1^1 < x_2 - x_2^1 < 0\},$$

$$III \stackrel{\triangle}{=} \{x \in R^2 | x_1 > x_1^1, -(x_1 - x_1^1) < x_2 - x_2^1 < 0\}.$$

For the initial state $x^0 \in II$ or III, bang–bang control is not feasible but constant control is. On the other hand, for $x^0 \in I$, bang–bang control is feasible but constant control is not except for x^0 belonging to the boundary of region I; however, in the latter situation, it is constant control of maximum magnitude and hence control with a single bang.

For $x^0 \in II$ or III the extremal control is unique and given by

$$u(t) \equiv \frac{x_2^1 - x_2^0}{x_1^1 - x_1^0} = \text{constant}. \qquad (12.11)$$

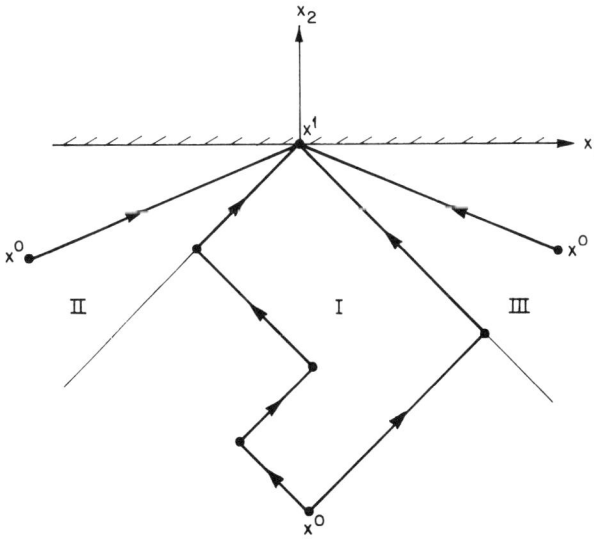

Figure 12.5. Initial state region.

On rewriting the cost as

$$t_1 - t_0 = \int_{x_2^0}^{x_2^1} \frac{dx_2(t)}{u^2(t)} \qquad (12.12)$$

by means of the second of (12.1), the cost due to use of control (12.11) is

$$t_1 - t_0 = \frac{(x_1^1 - x_1^0)^2}{x_2^1 - x_2^0}. \qquad (12.13)$$

For $x^0 \in I$ the extremal control is not unique; any feasible bang–bang control is extremal. The corresponding cost (12.12) is

$$t_1 - t_0 = x_2^1 - x_2^0. \qquad (12.14)$$

This is clearly a minimum and hence any extremal control is optimal for $x^0 \in I$. No such claim can be made for the unique extremal control (12.11) for $x^0 \in II$ or III; all we can assert is that it is a candidate for optimal control. However, we shall show subsequently (Section 15.5) that the extremal control is indeed optimal.

12.2. A Problem of Time-Optimal Navigation

The problem we are about to treat is a version of a classical navigation problem, that of steering a boat from one prescribed position to another one in minimum time; for instance, see Ref. 12.2.

Consider a boat moving with velocity **v** of constant magnitude v relative to a stream that flows with constant velocity **s** relative to the shore. Taking $v = 1$ (that is, normalizing with respect to the relative boat speed) and letting the x_1 and x_2 directions be parallel and normal, respectively, to the stream velocity **s**, the equations of motion of the boat (see Figure 12.6) are

$$\dot{x}_1(t) = s + u_1(t), \qquad \dot{x}_2(t) = u_2(t) \qquad (12.15)$$

where s is the constant magnitude of the stream velocity **s**, and $u_1 = \cos\phi$, $u_2 = \sin\phi$ with ϕ being the angle between **v** and **s**. Thus control u is constrained by

$$U = \{u \in R^2 \mid u_1^2 + u_2^2 = 1\}. \qquad (12.16)$$

Chap. 12 • Examples of Extremal Control

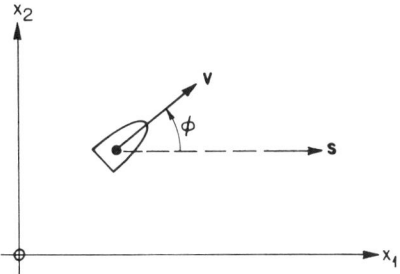

Figure 12.6. Time-optimal navigation.

The initial state, x^0, is prescribed as is the terminal one, x^1. We wish to determine the optimal steering program, that is, the control that minimizes the cost

$$\tau_1 \stackrel{\Delta}{=} t_1 - t_0 = \int_{t_0}^{t_1} dt. \qquad (12.17)$$

At this stage of the development of the subject we are again only able to deduce an *extremal* control. To utilize the maximum principle we require the function $H(\cdot)$; it is given by

$$H(\lambda, y, u) = \lambda_0 + \lambda_1(s + u_1) + \lambda_2 u_2. \qquad (12.18)$$

Then the adjoint equations are

$$\dot{\lambda}_0(t) = 0, \quad \dot{\lambda}_1(t) = 0, \quad \dot{\lambda}_2(t) = 0 \qquad (12.19)$$

so that

$$\lambda_0(t) \equiv \text{constant} \stackrel{\Delta}{=} \lambda_0,$$

$$\lambda_1(t) \equiv \text{constant} \stackrel{\Delta}{=} \lambda_1, \qquad (12.20)$$

$$\lambda_2(t) \equiv \text{constant} \stackrel{\Delta}{=} \lambda_2,$$

where, by (c) of Theorem 11.1, $\lambda_0 \leq 0$.

Now we invoke (a) of the theorem. In view of (12.18) this means that an extremal control $u(\cdot):[t_0, t_1] \to R^2$ is such that at each $t \in [t_0, t_1]$ its value,

$u(t)$, maximizes that portion of $H[\lambda(t), y(t), u]$ that depends on u, namely,

$$\lambda_1 u_1 + \lambda_2 u_2 = [\lambda_1 \; \lambda_2]\begin{bmatrix} u_1 \\ u_2 \end{bmatrix}.$$

Thus, for $i = 1, 2$, we obtain

$$\frac{u_i(t)}{[u_1^2(t) + u_2^2(t)]^{1/2}} = \frac{\lambda_i}{(\lambda_1^2 + \lambda_2^2)^{1/2}}$$

so that, as a consequence of the constraint (12.16),

$$u_i(t) = \frac{\lambda_i}{(\lambda_1^2 + \lambda_2^2)^{1/2}} \equiv \text{constant}, \quad i = 1, 2. \qquad (12.21)$$

In other words, a constant steering angle is extremal, and hence the velocity $\mathbf{v}$ relative to the stream as well as the velocity $\mathbf{s} + \mathbf{v}$ relative to the shore are constant. Consequently the path of the boat (as observed from the shore) due to an extremal control is rectilinear; see Figure 12.7. Therefore the corresponding transfer time, τ_1, must be such that, with $v = 1$,

$$(1 - s^2)\tau_1^2 + 2\Delta x_1 s \tau_1 - (\Delta x_1)^2 - (\Delta x_2)^2 = 0 \qquad (12.22)$$

where $\Delta x_1 \triangleq x_1^1 - x_1^0$ and $\Delta x_2 \triangleq x_2^1 - x_2^0$. Furthermore we have

$$u_1(t) \equiv \frac{\Delta x_1 - s\tau_1}{\tau_1}, \quad u_2(t) \equiv \frac{\Delta x_2}{\tau_1}. \qquad (12.23)$$

Now one must consider three possibilities:

(i) $s < 1$, (ii) $s = 1$, (iii) $s > 1$,

corresponding to the stream speed less than, equal to, and greater than the relative boat speed. In particular we are concerned with determining the region of initial states (positions), x^0, from which the given terminal position, x^1, can be reached by means of an extremal control; if an extremal control were indeed optimal, that would be the region $E^* \cup \theta^*$.

(i) $s < 1$. Upon solving (12.22) and invoking the requirement $\tau_1 \geq 0$, one has

$$\tau_1 = \frac{-s\Delta x_1 + [(\Delta x_1)^2 + (\Delta x_2)^2 - s^2(\Delta x_2)^2]^{1/2}}{1 - s^2}. \qquad (12.24)$$

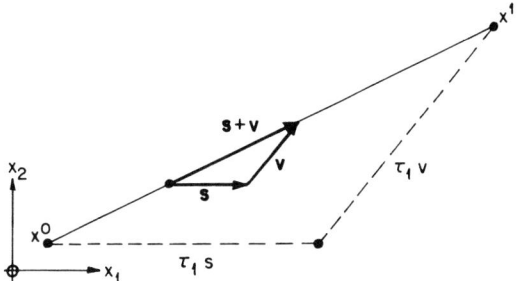

Figure 12.7. Trajectory generated by an extremal control.

Thus, for given Δx_1 and Δx_2, there exists a unique (nonnegative) transfer time τ_1; that is, x^1 can be reached from *all* points in R^2.

(ii) $s=1$. In this case the solution of (12.22) yields the unique solution

$$\tau_1 = \frac{(\Delta x_1)^2 + (\Delta x_2)^2}{2\Delta x_1}. \tag{12.25}$$

However, since $\tau_1 \in [0, \infty)$, it follows that $x^1 \neq x^0$ can be reached provided $\Delta x_1 > 0$; that is, the initial point region is the open half-plane together with the terminal point, namely, it is

$$\{x \in R^2 \mid x_1^0 < x_1^1\} \cup \{x^1\}.$$

(iii) $s > 1$. In this case, the solution of (12.22) has two roots; that is,

$$\tau_1 = \frac{s\Delta x_1 \pm \left[(\Delta x_1)^2 + (\Delta x_2)^2 - s^2(\Delta x_2)^2\right]^{1/2}}{s^2 - 1}. \tag{12.26}$$

Both roots are real if

$$(\Delta x_1)^2 + (\Delta x_2)^2 - s^2(\Delta x_2)^2 \geq 0. \tag{12.27}$$

It is established readily that the root corresponding to the plus sign in (12.26) is not due to an extremal control. To show this we invoke (b) and (c) of Theorem 11.1; namely,

$$H[\lambda(t), y(t), u(t)] = 0 \tag{12.28}$$

and

$$\lambda_0(t) \equiv \text{constant} \leq 0. \tag{12.29}$$

In view of (12.18) and (12.21) the condition (12.28) becomes

$$1 + su_1(t) + \frac{\lambda_0}{\left(\lambda_1^2 + \lambda_2^2\right)^{1/2}} = 0 \tag{12.30}$$

so that, with (12.29), we obtain

$$1 + su_1(t) \geq 0. \tag{12.31}$$

However, on substituting (12.23) for $u_1(t)$ and employing (12.26) with the plus sign, we arrive at

$$1 + su_1(t) = -\frac{\left[(\Delta x_1)^2 + (\Delta x_2)^2 - s^2(\Delta x_2)^2\right]^{1/2}}{\tau_1}, \tag{12.32}$$

which contradicts (12.31), except in the case of equality in (12.27) corresponding to the single root

$$\tau_1 = \frac{s\Delta x_1}{s^2 - 1}. \tag{12.33}$$

Thus, in general, only the root with the minus sign in (12.26) is due to an extremal control.

Now, as a consequence of the rectilinearity of a path generated by an extremal control and condition (12.27), such a path is subject to the condition

$$-(s^2 - 1)^{-1/2} \leq \frac{\Delta x_2}{\Delta x_1} \leq (s^2 - 1)^{-1/2}. \tag{12.34}$$

Also, because of the first of state equations (12.15) together with $s > 1$, $\tau_1 \geq 0$ and constraint (12.16), we can conclude that

$$\Delta x_1 \geq 0. \tag{12.35}$$

The conditions (12.34)–(12.35) define the initial state region.

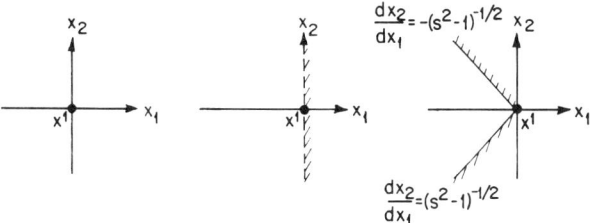

Figure 12.8. Initial state regions.

The regions of initial states (positions) from which an extremal control can steer the state to the given terminal one, x^1, are illustrated for cases (i)–(iii) in Figure 12.8.

It is noteworthy that in case (ii) the trajectories due to an extremal control cannot belong to the boundary of the initial state region; that is, one cannot reach x^1 in finite time if $x_1^0 = x_1^1$ and $x_2^0 \neq x_2^1$. However, in case (iii) the trajectories due to an extremal control can belong to the boundary of the initial state region; that is, x^0 can belong to the boundary. Finally, in case (iii) with x^0 and hence with the whole trajectory on the boundary, it follows from (12.30) with (12.32) and (12.34) that $\lambda_0(t) \equiv 0$; this is an example of so-called *abnormality* (for instance, see Ref. 12.3).

In conclusion, we found an extremal control for each case ($s < 1$, $s = 1$, $s > 1$). This control is unique; that is, for given x^0, x^1, and s there is only one control for which the necessary conditions of Theorem 11.1 are satisfied. On checking conditions which are sufficient (Chapter 15) it can be shown that the extremal control is indeed optimal (Exercise 15.13).

12.3. The Minimum Distance to a Given Curve

Now let us treat a simple example that serves to illustrate the use of the maximum principle including the terminal transversality condition. It is desired to determine the shape of the curve of minimum length that joins a given point and a point on a prescribed smooth curve lying in a plane; see Figure 12.9. With arc length as the independent variable, the state equations are

$$\dot{x}_1(t) = u_1(t), \qquad \dot{x}_2(t) = u_2(t) \tag{12.36}$$

where the control components, $u_i(t)$, are the direction cosines of the tangent

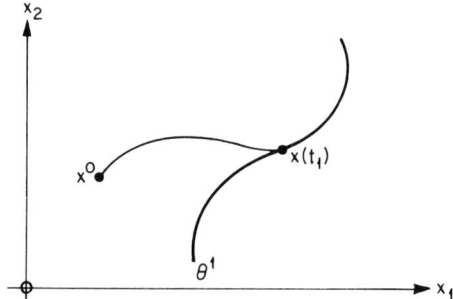

Figure 12.9. The minimum distance to a given curve.

to the curve; hence the control constraint set is

$$U = \{u \in R^2 \mid u_1^2 + u_2^2 = 1\}. \tag{12.37}$$

It is required to join the given initial point x^0 to the curve (target set)

$$\theta^1 = \{x \in R^2 \mid \theta^1(x) = 0\}, \tag{12.38}$$

where $\theta^1(\cdot): R^2 \to R^1$ is of class C^1 and grad $\theta^1(x) \neq 0$ for $x \in \theta^1$.

The cost is the total arc length

$$\tau_1 \stackrel{\triangle}{=} t_1 - t_0 = \int_{t_0}^{t_1} dt. \tag{12.39}$$

We begin again by deducing the extremal control(s), employing Theorem 11.1. The function $H(\cdot)$ is given by

$$H(\lambda, y, u) = \lambda_0 + \lambda_1 u_1 + \lambda_2 u_2$$

so that the adjoint equations for $j = 1, 2$ are

$$\dot{\lambda}_1(t) = 0, \quad \dot{\lambda}_2(t) = 0,$$

whence

$$\lambda_1(t) \equiv \text{constant} \stackrel{\triangle}{=} \lambda_1, \quad \lambda_2(t) \equiv \text{constant} \stackrel{\triangle}{=} \lambda_2. \tag{12.40}$$

Also, by condition (c), we have

$$\lambda_0(t) \equiv \text{constant} \stackrel{\Delta}{=} \lambda_0 \leq 0. \qquad (12.41)$$

As a consequence of the condition (a) and the constraint (12.37), an extremal control is such that

$$\begin{bmatrix} u_1(t) \\ u_2(t) \end{bmatrix} \quad \text{and} \quad \begin{bmatrix} \lambda_1 \\ \lambda_2 \end{bmatrix}$$

are codirectional; that is, an extremal control is such that

$$u_i(t) \equiv \frac{\lambda_i}{(\lambda_1^2 + \lambda_2^2)^{1/2}} = \text{constant}, \quad i = 1, 2. \qquad (12.42)$$

Thus, as expected, a curve due to an extremal control (slope) is a straight line with the slope

$$\frac{dx_2(t)}{dx_1(t)} = \frac{u_2(t)}{u_1(t)} \equiv \frac{\lambda_2}{\lambda_1} = \text{constant}. \qquad (12.43)$$

However, which straight line joining x^0 with a point on θ^1 is a candidate, or to put it another way, at what point of θ^1 can the line terminate? To decide this question we invoke the condition (d) of Theorem 11.1; namely, we impose (11.38) for η^1 satisfying (11.39). In particular we have

$$\lambda_1 \eta_1^1 + \lambda_2 \eta_2^1 = 0$$

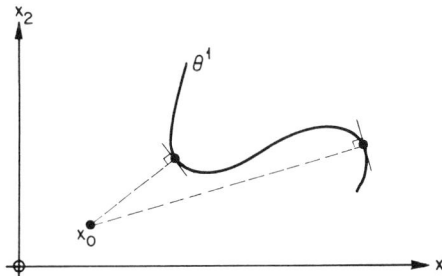

Figure 12.10. Candidates for curves of minimum length.

for all η_1^1, η_2^1 such that

$$\frac{\partial \theta^1(x(t_1))}{\partial x_1} \eta_1^1 + \frac{\partial \theta^1(x(t_1))}{\partial x_2} \eta_2^1 = 0,$$

whence

$$\lambda_2 \frac{\partial \theta^1(x(t_1))}{\partial x_1} - \lambda_1 \frac{\partial \theta^1(x(t_1))}{\partial x_2} = 0. \tag{12.44}$$

Then it follows from (12.43) with (12.44) that

$$\frac{x_2(t_1) - x_2^0}{x_1(t_1) - x_1^0} = \frac{\partial \theta^1(x(t_1))/\partial x_2}{\partial \theta^1(x(t_1))/\partial x_1}. \tag{12.45}$$

In other words, a candidate curve—that is, one that *may* be of minimum length—must be orthogonal to the given curve θ^1. Keep in mind that a straight line that emanates from the given point x^0, ends on θ^1, and is normal to it, need not be a curve of shortest length; this is illustrated in Figure 12.10.

Exercises

12.1. Suppose that the extremal controls deduced in Section 12.1 are indeed optimal. Find the corresponding limiting surfaces. Is the tangent plane defined at every point of a Σ-surface? Is every optimal trajectory regular? Discuss your answer.

12.2. Suppose that the extremal control deduced in Section 12.2 is indeed optimal. Find the corresponding limiting surfaces. Is the tangent plane defined at every point of a Σ-surface? Is every optimal trajectory regular? Discuss your answer.

12.3. In the problem of Section 12.3, let the target set $\theta^1 = \{x \in R^2 | x_1 = x_2\}$. Suppose that the extremal control is indeed optimal. Find the corresponding limiting surfaces. Is the tangent plane defined at every point of a Σ-surface? Is every optimal trajectory regular? Discuss your answer.

13

Some Generalizations

13.1. Introduction

In this chapter various generalizations of the optimal control problem will be discussed and corresponding necessary conditions for optimal control will be derived.

In the problem considered thus far the initial state is prescribed and the terminal state is required to belong to a given set, the target set θ^1. Our first generalization concerns a relaxation of the initial state specification to the less restrictive requirement that the initial state belong to a given set; that is, we seek controls that transfer the state from an unspecified point in a given set θ^0 to an unspecified point in a given set θ^1 while minimizing the cost.

In the type of problems treated heretofore the cost functional is an integral. However, in many problems of interest the cost depends directly on the terminal state. We shall consider controls which minimize the value of a given function of the terminal state.

Thus far the discussion has been restricted to problems with autonomous state equations as well as with cost integrands and end conditions which are not functions of the independent variable, that is, to autonomous systems. Now we shall admit nonautonomous state equations as well as cost integrands and end conditions which depend explicitly on t. In particular we shall consider problems in which the interval $[t_0, t_1]$ is prescribed.

The control constraint set, U, is a given subset of the control variable space, R^m; that is, it is a constant set. However, the control constraints may change as the state of the system changes. To account for this possibility we shall generalize our earlier discussion by allowing for the explicit dependence of the control constraint set on the state, x.

Next we shall treat the problem of optimal control in the presence of constraints which are not *pointwise*—that is, involve only the values of the control—but rather are constraints imposed on functionals. In particular we shall consider inequality constraints on integrals, the so-called *isoperimetric* constraints (such as the one encountered in Dido's problem).

The optimal control problem concerns the determination of *functions*, the control variables, so as to "steer" the state to the target set while rendering the minimum value of the cost functional. In many problems of interest the description of the system involves parameters, that is, constants, and some of these may be subject to selection by the designer of the system. In other words, the rate of change of the state may depend on the state, on the value of the control function, as well as on parameters. Finally we shall consider this possibility and discuss the optimal choice of the control function and of the parameter values.

13.2. The Initial Transversality Condition

Heretofore we have taken the initial state as a given point $x^0 \in R^n$. Now we wish to consider problems in which the initial state is merely restricted to belong to a given set $\theta^0 \subset R^n$. In particular we assume that the initial state set θ^0 is a nonempty smooth manifold of dimension $n-p$, $p \leq n$; namely,

$$\theta^0 \triangleq \{x \in R^n | \theta_i^0(x) = 0, i = 1, 2, \ldots, p\} \tag{13.1}$$

where the $\theta_i^0(\cdot): R^n \to R^1$ are given functions of class C^1, and the vectors grad $\theta_i^0(x), i = 1, 2, \ldots, p$, are linearly independent, or equivalently the matrix

$$\begin{bmatrix} \dfrac{\partial \theta_1^0(x)}{\partial x_1} & \cdots & \dfrac{\partial \theta_1^0(x)}{\partial x_n} \\ \vdots & & \vdots \\ \dfrac{\partial \theta_p^0(x)}{\partial x_1} & \cdots & \dfrac{\partial \theta_p^0(x)}{\partial x_n} \end{bmatrix}$$

has rank p at all $x \in \theta^0$; hence the tangent plane, $T_{\theta^0}(x)$, of θ^0 is defined at every $x \in \theta^0$.

Chap. 13 • Some Generalizations

We seek the *minimum minimorum* of the cost with respect to all feasible controls at a given initial state in θ^0 *and* with respect to all initial states in θ^0. In other words, for every $x^0 \in \theta^0 \cap E^*$ there is an optimal control yielding the minimum cost value $V^*(x^0)$. We seek that initial state and the corresponding optimal control yielding the minimum among the set of minimum cost values. Now we say that the control $u^*(\cdot):[t_0, t_1^*] \to R^m$, generating the solution $x^*(\cdot):[t_0, t_1^*] \to R^n$, $x^*(t_0) = x^{0*} \in \theta^0 \cap E^*$ *is optimal on θ^0* if and only if $u^*(\cdot)$ is optimal at x^{0*} and

$$V^*(x^{0*}) \leq V^*(x^0) \qquad \forall x^0 \in \theta^0 \cap [E^* \cup \theta^*]. \tag{13.2}$$

Of course, if θ^0 is a given point, $\theta^0 = \{x^0\}$, then "optimal on θ^0" and "optimal at x^0" are one and the same.

We proceed now in a manner similar to that employed in deriving the terminal transversality condition. Consider the initial state set in the augmented state space R^{n+1}, that is,

$$\Theta^0 \triangleq \{y \in R^{n+1} | x_0 = 0, x \in \theta^0\} \tag{13.3}$$

with the tangent plane at $y^0 \in \Theta^0$ given by

$$T_{\Theta^0}(y^0) = \{y \in R^{n+1} | x_0 = 0, x \in T_{\theta^0}(x^0)\}. \tag{13.4}$$

Next consider two states

$$y^{01} = \begin{bmatrix} 0 \\ \hline x^{01} \end{bmatrix} \quad \text{and} \quad y^{02} = \begin{bmatrix} 0 \\ \hline x^{02} \end{bmatrix}$$

in $\Theta^0 \cap [\mathcal{E}^* \cup \Theta^*]$. Now suppose that y^{01} is a *B*-point relative to $\Sigma(C_2)$, as illustrated in Figure 13.1; then it follows from (10.24) and (10.26) that

$$V^*(x^{01}) < V^*(x^{02}). \tag{13.5}$$

Let $\Gamma^*(C)$ denote the optimal trajectory generated by $u^*(\cdot)$ and starting at

$$y^*(t_0) = y^{0*} = \begin{bmatrix} 0 \\ \hline x^{0*} \end{bmatrix}.$$

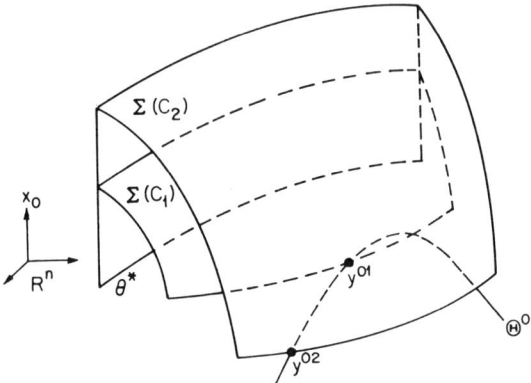

Figure 13.1. Initial state set.

Since $\Gamma^*(C)$ lies on $\Sigma(C)$, it follows from (13.2) and (13.5) that

$$\Theta^0 \cap B/\Sigma(C) = \emptyset. \tag{13.6}$$

If, as we assume, y^{0*} is a regular interior point of $\Sigma(C)$ so that its tangent plane $T_{\Sigma(C)}(y^{0*})$ is defined, then (13.6) implies that

$$T_{\Theta^0}(y^{0*}) \subset T_{\Sigma(C)}(y^{0*}). \tag{13.7}$$

Since $\lambda(t_0)$ is normal to $T_{\Sigma(C)}(y^{0*})$, it is also normal to $T_{\Theta^0}(y^{0*})$ and hence to every vector $\eta^0 \triangleq [\eta_0^0 \; \eta_1^0 \cdots \eta_n^0]^T$ in $T_{\Theta^0}(y^{0*})$; see Figure 13.2. Consequently we have the *initial transversality condition*

$$\lambda^T(t_0)\eta^0 = 0 \tag{13.8}$$

for all η^0 in $T_{\Theta^0}(y^{0*})$, that is, such that

$$\eta_0^0 = 0,$$

$$\sum_{j=1}^{n} \frac{\partial \theta_i^0(x^{0*})}{\partial x_j} \eta_j^0 = 0, \quad i = 1, 2, \ldots, p. \tag{13.9}$$

Equations (13.9) can be solved for $p+1$ components of η^0 as linear combinations of the remaining $n-p$ ones which are *arbitrary*. The coeffi-

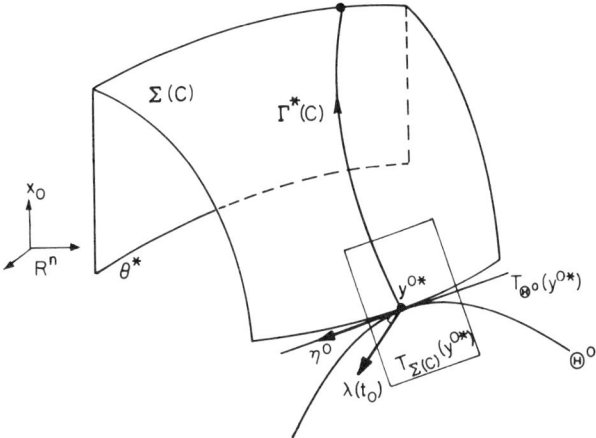

Figure 13.2. The initial transversality condition.

cients of the arbitrary components in (13.8) must vanish, yielding $n-p$ conditions involving $x^*(t_0) = x^{0*}$ and $\lambda(t_0)$; of course, $\lambda_0(t_0)$ does not appear in these relations. Thus the initial transversality condition and the initial conditions (13.1) provide n relations; together with the n terminal relations there are $2n$ end conditions.

If the initial state is prescribed, $\theta^0 = \{x^0\}$, so that

$$\theta_i^0(x) = x_i - x_i^0, \quad i = 1, 2, \ldots, p = n,$$

then the initial transversality condition is trivially satisfied.

If a control $u(\cdot):[t_0, t_1] \to R^m$ generates a solution $x(\cdot):[t_0, t_1] \to R^n$, $x(t_0) \in \theta^0$ and $x(t_1) \in \theta^1$, such that the necessary conditions embodied in Theorem 11.1 *and* in the initial transversality condition (13.8)–(13.9) are met, we still refer to it as an *extremal* control.

Before illustrating the use of the maximum principle, including the transversality conditions, we should point out that these conditions need not be satisfied—that is, are not necessary conditions for achieving a minimum value of the cost—unless $t_1^* > t_0$, as we have supposed. However, if $\theta^0 \cap \theta^1 \neq \emptyset$, the proviso that $t_1^* > t_0$ may be invalid. This is seen readily by considering an example: Determine the curve of minimum length between two *intersecting* curves, θ^0 and θ^1. Clearly it is the point of intersection.

13.3. Minimum Distance between Given Curves

Let there be given two nonintersecting plane curves

$$\theta^0 = \{x \in R^2 | \theta^0(x) = 0\}$$

and

$$\theta^1 = \{x \in R^2 | \theta^1(x) = 0\}.$$

What conditions must be met by the curve of minimum length joining θ^0 and θ^1? We have already discussed this question for the case in which θ^0 is a given point; see Section 12.3. We determined that such a curve must be a straight line that is orthogonal to the curve θ^1. Thus there remains the determination of the condition to be satisfied at the initial point $x(t_0) \in \theta^0$. The straightforward application of the initial transversality condition, together with (12.43), leads to

$$\frac{x_2(t_1) - x_2(t_0)}{x_1(t_1) - x_1(t_0)} = \frac{\partial \theta^0(x(t_0))/\partial x_2}{\partial \theta^0(x(t_0))/\partial x_1} \tag{13.10}$$

so that the curve of minimum length must be orthogonal to both θ^0 and θ^1.

For example, let

$$\theta^0(x) = (x_1 + a)^2 + x_2^2 - b^2$$

and

$$\theta^1(x) = (x_1 - a)^2 + x_2^2 - b^2$$

where $a = \text{constant} \neq 0$, $b^2 = \text{constant} < a^2$. Hence (12.45) and (13.10), together with (12.43), imply that

$$\frac{x_2(t)}{x_1(t) + a} = \frac{x_2(t)}{x_1(t) - a}$$

whence

$$x_2(t) \equiv 0.$$

13.4. A Terminal Cost

Here we are concerned with problems which differ in only one respect from those considered thus far; namely, the cost (10.4) is replaced by

$$V[x^0, u(\cdot), x(\cdot)] \triangleq \int_{t_0}^{t_1} f_0[x(t), u(t)] \, dt + G[x(t_1)] \quad (13.11)$$

where $G(\cdot): R^n \to R^1$ is a prescribed function of class C^1. Thus, in addition to an integral cost, there is a *terminal cost* $G[x(t_1)]$. Since the problem statement is unaltered in all other respects, we retain the definition of an "optimal control on θ^0" with the understanding that the cost, whose value is to be minimized, is defined by (13.11). We ask now: How are the necessary conditions for an optimal control affected by the presence of a terminal cost? In order to use the results already established, we introduce an augmented state

$$\tilde{x} \triangleq [x_1 \; x_2 \; \cdots \; x_{n+1}]^T = \begin{bmatrix} x \\ \hline x_{n+1} \end{bmatrix}$$

and control

$$\tilde{u} \triangleq [u_1 \; u_2 \; \cdots \; u_{m+1}]^T = \begin{bmatrix} u \\ \hline u_{m+1} \end{bmatrix},$$

and we consider the state equations

$$\dot{x}_j(t) = f_j[x(t), u(t)], \quad j = 1, 2, \ldots, n,$$
$$\dot{x}_{n+1}(t) = u_{m+1}(t) \quad (13.12)$$

with the end conditions

$$\tilde{x}(t_0) \in \tilde{\theta}^0 \triangleq \{\tilde{x} \in R^{n+1} | \tilde{\theta}_i^0(\tilde{x}) = 0, \; i = 1, 2, \ldots, p+1\}$$

where

$$\tilde{\theta}_i^0(\tilde{x}) = \theta_i^0(x), \quad i = 1, 2, \ldots, p,$$
$$\tilde{\theta}_{p+1}^0(\tilde{x}) = x_{n+1} \quad (13.13)$$

and

$$\tilde{x}(t_1) \in \tilde{\theta}^1 \triangleq \{\tilde{x} \in R^{n+1} | \tilde{\theta}_i^1(\tilde{x}) = 0, i = 1, 2, \ldots, q+1\}$$

where

$$\tilde{\theta}_i^1(\tilde{x}) = \theta_i^1(x), \qquad i = 1, 2, \ldots, q,$$
$$\tilde{\theta}_{q+1}^1(\tilde{x}) = x_{n+1} - G(x). \tag{13.14}$$

The control constraint set is now

$$\tilde{U} \triangleq \{\tilde{u} \in R^{m+1} | u \in U, |u_{m+1}| < \infty\}. \tag{13.15}$$

Finally the cost for the augmented system is

$$\int_{t_0}^{t_1} \{f_0[x(t), u(t)] + u_{m+1}(t)\} \, dt. \tag{13.16}$$

We claim the following: If $u^*(\cdot):[t_0, t_1^*] \to R^m$ is optimal on θ^0 for the *original* system with the cost (13.11) and generates the solution $x^*(\cdot):[t_0, t_1^*] \to R^n$, then the control

$$\tilde{u}^*(\cdot) = \begin{bmatrix} u^*(\cdot) \\ ----- \\ u_{m+1}^*(\cdot) \end{bmatrix} : [t_0, t_1^*] \to R^{m+1}$$

is optimal on $\tilde{\theta}^0$ for the augmented system with the cost (13.16) and generates the solution

$$\tilde{x}^*(\cdot) = \begin{bmatrix} x^*(\cdot) \\ ----- \\ x_{n+1}^*(\cdot) \end{bmatrix} : [t_0, t_1^*] \to R^{n+1},$$

where $u_{m+1}^*(\cdot)$ is any piecewise continuous function such that

$$\dot{x}_{n+1}^*(t) = u_{m+1}^*(t),$$
$$x_{n+1}^*(t_0) = 0, \qquad x_{n+1}^*(t_1^*) = G[x^*(t_1^*)].$$

Chap. 13 • Some Generalizations

Such a function exists; for instance, it is one defined by

$$u^*_{m+1}(t) \equiv \frac{G[x^*(t_1^*)]}{t_1^* - t_0}.$$

For suppose that $\tilde{u}^*(\cdot)$ is not optimal on $\tilde{\theta}^0$; then there is a

$$\tilde{u}(\cdot) = \begin{bmatrix} u(\cdot) \\ ----- \\ u_{m+1}(\cdot) \end{bmatrix} : [t_0, t_1] \to R^{m+1}$$

with the corresponding solution

$$\tilde{x}(\cdot) = \begin{bmatrix} x(\cdot) \\ ----- \\ x_{n+1}(\cdot) \end{bmatrix} : [t_0, t_1] \to R^{n+1}$$

of (13.12), satisfying (13.13)–(13.14) and such that

$$\int_{t_0}^{t_1} f_0[x(t), u(t)] \, dt + G[x(t_1)] < \int_{t_0}^{t_1^*} f_0[x^*(t), u^*(t)] \, dt + G[x^*(t_1^*)].$$

But that contradicts our assumption that $u^*(\cdot)$ is optimal on θ^0 for the original system.

Consequently *necessary* conditions for the optimality of $\tilde{u}^*(\cdot)$ on $\tilde{\theta}^0$ for the augmented system are also *necessary* for the optimality of $u^*(\cdot)$ on θ^0 for the original system. However, the cost (13.16) is an integral one; hence our earlier results—namely, Theorem 11.1 and the initial transversality condition of Section 13.2—apply. To utilize them we consider

$$\tilde{\lambda} \triangleq [\lambda_0 \ \lambda_1 \cdots \lambda_{n+1}]^T = \begin{bmatrix} \lambda \\ --- \\ \lambda_{n+1} \end{bmatrix},$$

$$\tilde{y} \triangleq [x_0 \ x_1 \cdots x_{n+1}]^T = \begin{bmatrix} y \\ --- \\ y_{n+1} \end{bmatrix},$$

and

$$\tilde{H}(\tilde{\lambda}, \tilde{y}, \tilde{u}) \triangleq H(\lambda, y, u) + \lambda_0 u_{m+1} + \lambda_{n+1} u_{m+1}. \tag{13.17}$$

Thus the adjoint equations are

$$\dot{\lambda}_0(t) = 0,$$

$$\dot{\lambda}_j(t) = -\frac{\partial H[\lambda(t), y^*(t), u^*(t)]}{\partial x_j}, \quad j=1,2,\ldots,n, \quad (13.18)$$

$$\dot{\lambda}_{n+1}(t) = 0.$$

The conditions (a) and (b) of Theorem 11.1 become

$$\max_{\tilde{u} \in \tilde{U}} \tilde{H}[\tilde{\lambda}(t), \tilde{y}^*(t), \tilde{u}] = \tilde{H}[\tilde{\lambda}(t), \tilde{y}^*(t), \tilde{u}^*(t)] = 0. \quad (13.19)$$

Since $|u^*_{m+1}(t)| < \infty$, it follows from (13.19) that

$$\frac{\partial \tilde{H}[\tilde{\lambda}(t), \tilde{y}^*(t), \tilde{u}^*(t)]}{\partial u_{m+1}} = 0$$

so that, in view of (13.17),

$$\lambda_0(t) + \lambda_{n+1}(t) = 0.$$

Hence, as a consequence of (13.18) and the condition (c) of Theorem 11.1, we have

$$\lambda_{n+1}(t) = -\lambda_0(t) \equiv \text{constant} \stackrel{\Delta}{=} -\lambda_0 \geq 0. \quad (13.20)$$

Now it follows from $\tilde{\lambda}(t) \neq 0$, together with (13.20), that $\lambda(t) \neq 0$.
As a result of (13.17) with (13.15) and (13.20), the condition (13.19) reduces to

$$\max_{u \in U} H[\lambda(t), y^*(t), u] = H[\lambda(t), y^*(t), u^*(t)] = 0. \quad (13.21)$$

Finally we turn to the transversality conditions. The initial one is

$$\sum_{j=1}^{n+1} \lambda_j(t_0) \eta_j^0 = 0$$

subject to

$$\sum_{j=1}^{n} \frac{\partial \tilde{\theta}_i^0(\tilde{x}^*(t_0))}{\partial x_j} \eta_j^0 = 0, \quad i=1,2,\ldots,p,$$

$$\eta_{n+1}^0 = 0,$$

so that

$$\sum_{j=1}^{n} \lambda_j(t_0) \eta_j^0 = 0 \quad (13.22)$$

subject to

$$\sum_{j=1}^{n} \frac{\partial \theta_i^0(x^*(t_0))}{\partial x_j} \eta_j^0 = 0, \quad i=1,2,\ldots,p. \quad (13.23)$$

The terminal one is

$$\sum_{j=1}^{n+1} \lambda_j(t_1^*) \eta_j^1 = 0$$

subject to

$$\sum_{j=1}^{n} \frac{\partial \tilde{\theta}_i^1(\tilde{x}^*(t_1^*))}{\partial x_j} \eta_j^1 = 0, \quad i=1,2,\ldots,q,$$

$$\eta_{n+1}^1 - \sum_{j=1}^{n} \frac{\partial G[x^*(t_1^*)]}{\partial x_j} \eta_j^1 = 0,$$

whence, upon the use of (13.20), we obtain

$$\sum_{j=1}^{n} \left\{ -\lambda_0 \frac{\partial G[x^*(t_1^*)]}{\partial x_j} + \lambda_j(t_1^*) \right\} \eta_j^1 = 0 \quad (13.24)$$

subject to

$$\sum_{j=1}^{n} \frac{\partial \theta_i^1(x^*(t_1^*))}{\partial x_j} \eta_j^1 = 0, \quad i=1,2,\ldots,q. \quad (13.25)$$

In summary, the necessary conditions for the optimality of $\tilde{u}^*(\cdot)$ on $\tilde{\theta}^0$, and hence for the optimality of $u^*(\cdot)$ on θ^0, reduce to condition (13.21) with the adjoint equations (13.18) and the transversality conditions (13.22)–(13.23) and (13.24)–(13.25). These conditions are devoid of λ_{n+1} and u_{m+1}. Indeed the only change due to the presence of the terminal cost is an added term in the terminal transversality condition (13.24)–(13.25), as can be seen by comparison with (11.38)–(11.39).

While it is not required for the purpose of deriving *necessary* conditions, the converse of our claim is also true; that is, if

$$\tilde{u}^*(\cdot) = \begin{bmatrix} u^*(\cdot) \\ ----- \\ u^*_{m+1}(\cdot) \end{bmatrix}$$

is optimal on $\tilde{\theta}^0$ for the augmented system with the cost (13.16), then $u^*(\cdot)$ is optimal on θ^0 for the original system with the cost (13.11). The proof is left to the reader, see Exercise 13.2.

Finally we note that the situation is much simpler if the initial state is prescribed, say $x(t_0) = x^0$. In that event one can pose a problem with an integral cost that is entirely equivalent to the original one with the cost (13.11). To do so we write

$$G[x(t_1)] = \int_{t_0}^{t_1} \mathrm{grad}^T G[x(t)] f[x(t), u(t)] \, dt + G(x^0).$$

Since the term $G(x^0)$ is constant for all controls one can replace (13.11) by the integral cost

$$\int_{t_0}^{t_1} \{f_0[x(t), u(t)] + \mathrm{grad}^T G[x(t)] f[x(t), u(t)]\} \, dt$$

without affecting the choice of an optimal control.

13.5. An Extremal Thrust Control for a Rocket

The system under consideration here is a thrust-limited rocket moving in a vertical plane under the influence of a constant gravitational attraction per unit mass.

Let t be the time, x_1, x_2 be the position coordinates, x_3, x_4 be the velocity components, x_5 be the mass, u_1, u_2 be the direction cosines of the

Chap. 13 • Some Generalizations

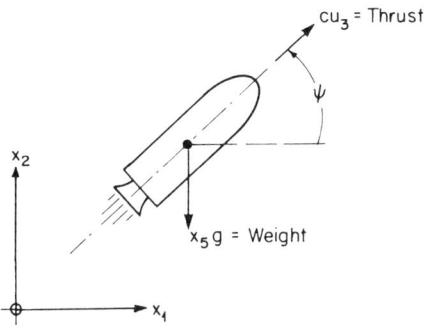

Figure 13.3. Force system acting on a rocket.

thrust, u_3 be the mass flow rate, c be the exhaust speed = const > 0, and g be the gravitational attraction per unit mass = const > 0.

The state equations are (see Ref. 13.1 and Figure 13.3)

$$\dot{x}_1(t) = x_3(t),$$
$$\dot{x}_2(t) = x_4(t),$$
$$\dot{x}_3(t) = \frac{c}{x_5(t)} u_1(t) u_3(t), \qquad (13.26)$$
$$\dot{x}_4(t) = \frac{c}{x_5(t)} u_2(t) u_3(t) - g,$$
$$\dot{x}_5(t) = -u_3(t).$$

The thrust magnitude, cu_3, is limited; namely, the mass flow rate is constrained by

$$0 \leq u_3(t) \leq u_3^{\max} \qquad (13.27)$$

where $u_3^{\max}$ is a prescribed positive constant. The direction cosines of the thrust direction satisfy the constraint

$$u_1^2(t) + u_2^2(t) = 1. \qquad (13.28)$$

The relations (13.27) and (13.28) define the control constraint set U.

It is desired to transfer the rocket so that given end conditions are met; that is, with

$$x \triangleq [x_1 \ x_2 \ \cdots \ x_5]^T$$

we impose

$$\theta_i^0(x(t_0))=0, \quad i=1,2,\ldots,p\leq 5,$$
$$\theta_i^1(x(t_1))=0, \quad i=1,2,\ldots,q\leq 5.$$

Such end conditions might specify the initial and terminal values of the rocket mass and hence the amount of fuel, of the position, of the velocity, or of combinations of these.

Furthermore it is desired to vary the rocket thrust in magnitude and direction so as to effect the transfer while minimizing a terminal cost $G[x(t_1)]$. Included are problems of minimizing fuel consumption, maximizing payload, maximizing terminal energy, and so on. A class of such problems is discussed in Refs. 13.1 and 13.2.

To deduce an extremal control we utilize the necessary conditions as modified for a terminal cost in Section 13.4.

For a purely terminal cost, $f_0(x,u)\equiv 0$ so that

$$H(\lambda,y,u)=\lambda_1 x_3+\lambda_2 x_4-\lambda_4 g+\left[\frac{c}{x_5}(\lambda_3 u_1+\lambda_4 u_2)-\lambda_5\right]u_3. \quad (13.29)$$

Thus, if $u(\cdot):[t_0,t_1]\to R^3$ is an extremal control that generates the solution $x(\cdot):[t_0,t_1]\to R^5$, the adjoint equations for $j=1,2,\ldots,5$ are

$$\dot{\lambda}_1(t)=0,$$
$$\dot{\lambda}_2(t)=0,$$
$$\dot{\lambda}_3(t)=-\lambda_1(t), \quad (13.30)$$
$$\dot{\lambda}_4(t)=-\lambda_2(t),$$
$$\dot{\lambda}_5(t)=\frac{c}{x_5^2(t)}[\lambda_3(t)u_1(t)+\lambda_4(t)u_2(t)]u_3(t).$$

The first four of these are integrated readily; namely,

$$\lambda_1(t)=\lambda_1(t_1),$$
$$\lambda_2(t)=\lambda_2(t_1),$$
$$\lambda_3(t)=\lambda_3(t_1)+\lambda_1(t_1)(t_1-t),$$
$$\lambda_4(t)=\lambda_4(t_1)+\lambda_2(t_1)(t_1-t).$$

Let us consider first the extremal thrust direction, $u_1(t)$ and $u_2(t)$. It is only defined when there is a thrust, that is, when $u_3(t) > 0$. Then, as a consequence of the condition (a) of Theorem 11.1 with $x_5(t) > 0$ (as in all problems of interest; see Sec. 14.9), together with the constraint (13.28), we have

$$u_i(t) = \frac{\lambda_{i+2}(t)}{[\lambda_3^2(t) + \lambda_4^2(t)]^{1/2}}, \quad i = 1, 2. \tag{13.31}$$

We conclude that the thrust direction angle

$$\psi(t) = \tan^{-1} \frac{u_2(t)}{u_1(t)} = \tan^{-1} \frac{\lambda_4(t)}{\lambda_3(t)}$$

varies continuously with the possible exception of a *single* discontinuity on $[t_0, t_1]$, unless $\lambda_i(t) \equiv 0$, $i = 1, 2, \ldots, 4$, and $u_3(t) \equiv 0$, or $\lambda_i(t) \equiv 0$, $i = 1, 2, \ldots, 5$, so that the maximum principle yields no information (Exercise 13.4). This discontinuity can occur if $\lambda_3(\cdot)$ and $\lambda_4(\cdot)$ pass through zero simultaneously and hence change sign; in that event the thrust direction experiences a jump of π radians.

Now let us turn to the consideration of the extremal thrust magnitude as determined by the mass flow rate, $u_3(\cdot)$. By the condition (a) of Theorem 11.1 and the constraint (13.27), the value of $u_3(\cdot)$ depends on the so-called *switching function*, $\sigma(\cdot): [t_0, t_1] \to R^1$; in view of (13.29) it is defined by

$$\sigma(t) \triangleq \frac{c}{x_5(t)} [\lambda_3(t) u_1(t) + \lambda_4(t) u_2(t)] - \lambda_5(t).$$

As a consequence of (13.28) and (13.31) this becomes

$$\sigma(t) = \frac{c}{x_5(t)} [\lambda_3^2(t) + \lambda_4^2(t)]^{1/2} - \lambda_5(t). \tag{13.32}$$

The term switching function is used because

$$\begin{aligned} u_3(t) &= 0 & \text{if } \sigma(t) < 0, \\ u_3(t) &= u_3^{\max} & \text{if } \sigma(t) > 0. \end{aligned} \tag{13.33}$$

Thus we need to investigate $\sigma(\cdot)$. We do this by considering $\dot{\sigma}(t)$. In view of (13.30) and (13.31) we obtain

$$\dot{\sigma}(t) = -\frac{c}{x_5(t)} [\lambda_1(t_1) u_1(t) + \lambda_2(t_1) u_2(t)]. \tag{13.34}$$

If $\lambda_1(t_1) \neq 0$ and $\lambda_2(t_1) \neq 0$, then $\dot{\sigma}(\cdot)$ can have at most one zero on $[t_0, t_1]$. It follows from (13.33) that $u_3(\cdot)$ is *bang-bang*; namely, either the thrust is at the maximum value or it is zero.

If $\lambda_1(t_1) \neq 0$ and $\lambda_2(t_1) = 0$, or conversely, then $\dot{\sigma}(t) = 0$ implies that $u_1(t) = 0$ or $u_2(t) = 0$. However, this can occur at most once on $[t_0, t_1]$, namely, at

$$t = t_1 + \frac{\lambda_3(t_1)}{\lambda_1(t_1)} \quad \text{or} \quad t = t_1 + \frac{\lambda_4(t_1)}{\lambda_2(t_1)}.$$

Hence, $\sigma(\cdot)$ cannot be zero on an interval and so $u_3(\cdot)$ is again bang–bang.

If $\lambda_1(t_1) = \lambda_2(t_1) = 0$, then $\dot{\sigma}(t) \equiv 0$ so that $\sigma(t) \equiv \text{constant}$. There are two possibilities. If

$$\sigma(t) \equiv \text{constant} \neq 0,$$

then $u_3(\cdot)$ does not switch; that is,

$$u_3(t) \equiv 0 \quad \text{or} \quad u_3(t) \equiv u_3^{\max}.$$

On the other hand, if

$$\sigma(t) \equiv 0,$$

then the maximum principle furnishes no information concerning $u_3(\cdot)$. We expect this to be the case if the cost does not depend on the choice of $u_3(\cdot)$; for example, see Exercise 13.5. From the condition (b) of Theorem 11.1, namely,

$$H[\lambda(t), y(t), u(t)] = 0,$$

it is readily seen that this exceptional case corresponds to

$$\lambda_3(t) = \lambda_3(t_1),$$
$$\lambda_4(t) = 0,$$
$$\lambda_5(t) = c \frac{\sqrt{\lambda_3^2(t_1)}}{x_5(t)},$$

and hence to

$$u_1(t) \equiv \text{sgn } \lambda_3(t_1), \quad u_2(t) \equiv 0.$$

Except for the cases in which the maximum principle fails to furnish information about $u_3(\cdot)$, one can show easily that $u_3(\cdot)$ is bang–bang with at most two switches. If $\lambda_3(\cdot)$ and $\lambda_4(\cdot)$ do not pass through zero simultaneously, then $\dot{\sigma}(\cdot)$ is continuous and can have at most one zero. If $\lambda_3(\cdot)$ and $\lambda_4(\cdot)$ pass through zero simultaneously, then $\dot{\sigma}(\cdot)$ experiences a discontinuity; however, this can occur but once. Thus, in either case, the switching function can have at most two zeros. It can also be shown (see Ref. 13.3 and Exercise 13.6) that there are only five possibilities:

(i) $u_3(t) = 0$ for $t \in [t_0, t_1]$,

(ii) $u_3(t) = u_3^{\max}$ for $t \in [t_0, t_1]$,

(iii) $u_3(t) = 0$ for $t \in [t_0, t')$,

       $u_3(t) = u_3^{\max}$ for $t \in [t', t_1]$,

(iv) $u_3(t) = u_3^{\max}$ for $t \in [t_0, t')$,

       $u_3(t) = 0$ for $t \in [t', t_1]$,

(v) $u_3(t) = u_3^{\max}$ for $t \in [t_0, t')$,

       $u_3(t) = 0$ for $t \in [t', t'')$,

       $u_3(t) = u_3^{\max}$ for $t \in [t'', t_1]$.

13.6. Nonautonomous Systems

Thus far the discussion has been restricted to systems for which the state equations, the cost integrand, and the end conditions do not depend explicitly on the independent variable; that is, the functions $f_i(\cdot, u)$, $i = 0, 1, \ldots, n$, as well as the $\theta_i^0(\cdot), i = 1, 2, \ldots, p$, and $\theta_i^1(\cdot), i = 1, 2, \ldots, q$, are functions from the *state* space R^n into R^1. Now we wish to allow for the explicit dependence on t. We dub such systems *nonautonomous*; the state equations are now

$$\dot{x}_i(t) = \hat{f}_i[x(t), t, u(t)], \qquad i = 1, 2, \ldots, n, \qquad (13.35)$$

where the $\hat{f}_i(\cdot): R^n \times R^1 \times R^m \to R^1$ are given functions such that the $\hat{f}_i(\cdot)$, $\partial \hat{f}_i(\cdot)/\partial x_j$ and $\partial \hat{f}_i(\cdot)/\partial t, i, j = 1, 2, \ldots, n$, are continuous, and the function $x(\cdot): [t_0, t_1] \to R^n$, $x(t_0) = x^0$, is a solution of (13.35) generated by an admissible control $u(\cdot): [t_0, t_1] \to R^m$.

As before we impose end conditions. Now, however, the initial conditions are

$$\hat{\theta}_i^0(x(t_0), t_0) = 0, \quad i = 1, 2, \ldots, p+1 \leq n+1, \tag{13.36}$$

and the terminal conditions are

$$\hat{\theta}_i^1(x(t_1), t_1) = 0, \quad i = 1, 2, \ldots, q+1 \leq n+1, \tag{13.37}$$

where the $\hat{\theta}_i^0(\cdot): R^n \times R^1 \to R^1$ and the $\hat{\theta}_i^1(\cdot): R^n \times R^1 \to R^1$ are given functions of class C^1.

The cost, to be minimized, is now defined by

$$\hat{V}[x^0, t_0, u(\cdot), x(\cdot)] \triangleq \int_{t_0}^{t_1} \hat{f}_0[x(t), t, u(t)] \, dt \tag{13.38}$$

where $\hat{f}_0(\cdot): R^n \times R^1 \times R^m \to R^1$ is a prescribed function of the same class as the $\hat{f}_i(\cdot)$.

We retain the definition of an admissible control $u(\cdot)$; i.e., $u(\cdot): [t_0, t_1] \to R^m$ is piecewise continuous and

$$u(t) \in U$$

where $U \subset R^m$ is given.

The explicit dependence on t results in a salient difference between the system (13.35)–(13.38) and the one treated thus far; namely, the translation property of the interval discussed in Section 10.2 is no longer valid. In other words, the initial value of t cannot be assigned arbitrarily. However, in all other respects we are able to "translate" the nonautonomous system into an "equivalent" one of the type considered heretofore. We do this by enlarging the state space to R^{n+1}, that is, by augmenting the state to

$$\tilde{x} \triangleq [x_1 \ x_2 \ \ldots \ x_{n+1}]^T = \begin{bmatrix} x \\ \hline x_{n+1} \end{bmatrix}.$$

The added state variable $x_{n+1}(\cdot): [t_0, t_1] \to R^1$ is given by

$$x_{n+1}(t) = t. \tag{13.39}$$

In other words, we introduce the new state space R^{n+1} as the product of the original one, R^n, with that of the independent variable, R^1.

Chap. 13 • Some Generalizations

Now let the $\tilde{f}_i(\cdot): R^{n+1} \times R^m \to R^1, i=1,2,\ldots,n+1$, be such that

$$\tilde{f}_i(\tilde{x}, u) = \hat{f}_i(x, x_{n+1}, u), \quad i=1,2,\ldots,n,$$

$$\tilde{f}_{n+1}(\tilde{x}, u) = 1.$$

Thus the new state equations are

$$\begin{aligned}\dot{x}_i(t) &= \tilde{f}_i[\tilde{x}(t), u(t)], \quad i=1,2,\ldots,n, \\ \dot{x}_{n+1}(t) &= 1.\end{aligned} \qquad (13.40)$$

Further let $\tilde{\theta}_i^0(\cdot): R^{n+1} \to R^1, i=1,2,\ldots,p+1$, and $\tilde{\theta}_i^1(\cdot): R^{n+1} \to R^1$, $i=1,2,\ldots,q+1$, be such that

$$\tilde{\theta}_i^0(\tilde{x}) = \hat{\theta}_i^0(x, x_{n+1}),$$

$$\tilde{\theta}_i^1(\tilde{x}) = \hat{\theta}_i^1(x, x_{n+1}).$$

Then the new end conditions are

$$\begin{aligned}\tilde{x}(t_0) \in \tilde{\theta}^0 &\triangleq \{\tilde{x} \in R^{n+1} \mid \tilde{\theta}_i^0(\tilde{x}) = 0, i=1,2,\ldots,p+1\}, \\ \tilde{x}(t_1) \in \tilde{\theta}^1 &\triangleq \{\tilde{x} \in R^{n+1} \mid \tilde{\theta}_i^1(\tilde{x}) = 0, i=1,2,\ldots,q+1\}.\end{aligned} \qquad (13.41)$$

To conform to our earlier assumptions we suppose that the grad $\tilde{\theta}_i^0(\tilde{x})$, $i=1,2,\ldots,p+1$, are linearly independent at every $\tilde{x} \in \tilde{\theta}^0$, and the grad $\tilde{\theta}_i^1(\tilde{x})$, $i=1,2,\ldots,q+1$, are linearly independent at every $\tilde{x} \in \tilde{\theta}^1$.

Finally, with $\tilde{f}_0(\tilde{x}, u) = \hat{f}_0(x, x_{n+1}, u)$, the new cost is defined by

$$\tilde{V}[\tilde{x}^0, u(\cdot), \tilde{x}(\cdot)] \triangleq \int_{t_0}^{t_1} \tilde{f}_0[\tilde{x}(t), u(t)] \, dt. \qquad (13.42)$$

The systems (13.35)–(13.38) and (13.40)–(13.42) are "equivalent" in the following sense. Let $u(\cdot): [t_0, t_1] \to R^m$ be an admissible control. Given x^0 and t_0, let $x(\cdot): [t_0, t_1] \to R^n$, $x(t_0) = x^0$, be the solution generated by $u(\cdot)$. Then $u(\cdot)$ generates the solution $\tilde{x}(\cdot): [t_0, t_1] \to R^{n+1}$,

$$\tilde{x}(t_0) = \tilde{x}^0 = \begin{bmatrix} x^0 \\ x_{n+1}^0 \end{bmatrix}$$

with $x_{n+1}^0 = t_0$, of (13.40), where

$$\tilde{x}(t) = \begin{bmatrix} x(t) \\ \hdashline x_{n+1}(t) \end{bmatrix}$$

with $x_{n+1}(t) = t$. Furthermore then we have

$$\tilde{V}[\tilde{x}^0, u(\cdot), \tilde{x}(\cdot)] = \hat{V}[x^0, t_0, u(\cdot), x(\cdot)].$$

Finally the end conditions imposed on the systems, namely, (13.36)–(13.37) and (13.41), respectively, are one and the same. Thus, in seeking an admissible control that results in a solution which satisfies the prescribed end conditions and results in the minimum value of the cost, one can consider the system (13.40)–(13.42) in place of (13.35)–(13.38). However, the former system is one of the type treated heretofore; in particular it is one to which Theorem 11.1 and Section 13.2 apply.

Let

$$\tilde{y} \triangleq [x_0 \ x_1 \cdots x_{n+1}]^T = \begin{bmatrix} y \\ \hdashline x_{n+1} \end{bmatrix},$$

$$\tilde{\lambda} \triangleq [\lambda_0 \ \lambda_1 \cdots \lambda_{n+1}]^T = \begin{bmatrix} \lambda \\ \hdashline \lambda_{n+1} \end{bmatrix},$$

and

$$\tilde{H}(\tilde{\lambda}, \tilde{y}, u) \triangleq \hat{H}(\lambda, y, t, u) + \lambda_{n+1}$$

where

$$\hat{H}(\lambda, y, t, u) \triangleq \sum_{i=0}^{n} \lambda_i f_i(x, t, u). \tag{13.43}$$

Then the adjoint equations are

$$\dot{\lambda}_0(t) = 0,$$

$$\dot{\lambda}_i(t) = -\frac{\partial \tilde{H}[\tilde{\lambda}(t), \tilde{y}^*(t), u^*(t)]}{\partial x_i}, \qquad i = 1, 2, \ldots, n+1. \tag{13.44}$$

The conditions (a) and (b) of Theorem 11.1 are

$$\max_{u \in U} \tilde{H}[\tilde{\lambda}(t), \tilde{y}^*(t), u] = \tilde{H}[\tilde{\lambda}(t), \tilde{y}^*(t), u^*(t)] = 0.$$

Chap. 13 • Some Generalizations

In view of (13.43), this condition reduces to

$$\max_{u \in U} \hat{H}[\lambda(t), y^*(t), t, u] = \hat{H}[\lambda(t), y^*(t), t, u^*(t)] \quad (13.45)$$

and

$$\hat{H}[\lambda(t), y^*(t), t, u^*(t)] = -\lambda_{n+1}(t). \quad (13.46)$$

Now it follows from (13.46) with (13.39) and (13.44) that

$$\hat{H}[\lambda(t), y^*(t), t, u^*(t)] = \hat{H}[\lambda(t_0^*), y^*(t_0^*), t_0^*, u^*(t_0^*)]$$

$$+ \int_{t_0^*}^t \frac{\partial \hat{H}[\lambda(\tau), y^*(\tau), \tau, u^*(\tau)]}{\partial \tau} d\tau. \quad (13.47)$$

Furthermore the adjoint equations (13.44) can be replaced by

$$\dot{\lambda}_0(t) = 0,$$
$$\dot{\lambda}_i(t) = -\frac{\partial \hat{H}[\lambda(t), y^*(t), t, u^*(t)]}{\partial x_i}, \quad i = 1, 2, \ldots, n, \quad (13.48)$$

since we have no further use for $\lambda_{n+1}(t)$. Also, in view of (13.46), $\tilde{\lambda}(t) \neq 0$ implies that $\lambda(t) \neq 0$.

The initial transversality condition is

$$\sum_{i=1}^n \lambda_i(t_0^*) \eta_i^0 + \lambda_{n+1}(t_0^*) \eta_{n+1}^0 = 0$$

subject to

$$\sum_{i=1}^n \frac{\partial \tilde{\theta}_j^0(\tilde{x}^*(t_0^*))}{\partial x_i} \eta_i^0 + \frac{\partial \tilde{\theta}_j^0(\tilde{x}^*(t_0^*))}{\partial x_{n+1}} \eta_{n+1}^0 = 0, \quad j = 1, 2, \ldots, p+1.$$

Upon invoking (13.39) and (13.46) these relations reduce to

$$\sum_{i=1}^n \lambda_i(t_0^*) \eta_i^0 = \hat{H}[\lambda(t_0^*), y^*(t_0^*), t_0^*, u^*(t_0^*)] \eta_{n+1}^0 \quad (13.49)$$

subject to

$$\sum_{i=1}^{n} \frac{\partial \hat{\theta}_j^0(x^*(t_0^*), t_0^*)}{\partial x_i} \eta_i^0 = -\frac{\partial \hat{\theta}_j^0(x^*(t_0^*), t_0^*)}{\partial t} \eta_{n+1}^0, \quad j=1,2,\ldots, p+1.$$

(13.50)

Similarly the terminal transversality condition becomes

$$\sum_{i=1}^{n} \lambda_i(t_1^*) \eta_i^1 = \hat{H}[\lambda(t_1^*), y^*(t_1^*), t_1^*, u^*(t_1^*)] \eta_{n+1}^1 \qquad (13.51)$$

subject to

$$\sum_{i=1}^{n} \frac{\partial \hat{\theta}_j^1(x^*(t_1^*), t_1^*)}{\partial x_i} \eta_i^1 = -\frac{\partial \hat{\theta}_j^1(x^*(t_1^*), t_1^*)}{\partial t} \eta_{n+1}^1, \quad j=1,2,\ldots, q+1.$$

(13.52)

We conclude that the conditions (a) and (b) of Theorem 11.1 are replaced by (13.45) and (13.47), respectively, while the condition (c) remains unchanged. The transversality conditions (11.38)–(11.39) and (13.8)–(13.9) are replaced by (13.49)–(13.50) and (13.51)–(13.52), respectively. The adjoint equations (11.28) are replaced by (13.48). Finally we note that none of these relations involves x_{n+1} or λ_{n+1}.

13.7. Problems with a Fixed Interval

One type of problem of interest is that of the optimal control of a system for which the interval $[t_0, t_1]$ is prescribed, but which is otherwise of the kind treated in Chapters 10 and 11. Thus it is a problem that constitutes a special case of those considered in Section 13.6.

The interval $[t_0, t_1]$ is fixed by prescribing the values of t_0 and t_1. The end conditions (13.36) and (13.37) reduce to

$$\hat{\theta}_i^0(x(t_0), t_0) = \theta_i^0(x(t_0)) = 0, \quad i=1,2,\ldots, p,$$
$$\hat{\theta}_{p+1}^0(x(t_0), t_0) = t_0 - a = 0, \quad a = \text{constant},$$

(13.53)

and

$$\hat{\theta}_i^1(x(t_1), t_1) = \theta_i^1(x(t_1)) = 0, \quad i = 1, 2, \ldots, q$$
$$\hat{\theta}_{q+1}^1(x(t_1), t_1) = t_1 - b = 0, \quad b = \text{constant.}$$
(13.54)

Since the state equations are autonomous and the cost integrand does not depend explicitly on t, $\hat{H}(\cdot)$ reduces to $H(\cdot)$; in particular, the condition (13.45) reduces to (a) of Theorem 11.1, while (13.47) becomes

$$H[\lambda(t), y^*(t), u^*(t)] = H[\lambda(a), y^*(a), u^*(a)] \quad (13.55)$$

replacing (b) of Theorem 11.1. Furthermore the adjoint equations and the transversality conditions revert to their original forms (11.28), (11.38)–(11.39), and (13.8)–(13.9), respectively. Thus the only change due to the prescription of the interval $[t_0, t_1]$ is the replacement of the condition (b) of Theorem 11.1 by (13.55). The knowledge of $H[\lambda(t_0), y^*(t_0), u^*(t_0)]$, namely, zero, is replaced by the knowledge of the value of $t_1 - t_0$.

13.8. A Minimum Fuel Rendezvous of a Constant-Power Rocket

Consider again the system discussed in Section 12.1, namely, a rocket operating at constant propulsive power. However, now we prescribe the initial and terminal values of the position and velocity as well as of the transfer time, and we seek a thrust-acceleration program that results in minimum fuel consumption. The equations for the system are

$$\dot{x}_1(t) = x_2(t),$$
$$\dot{x}_2(t) = u(t)$$
(13.56)

where t is time, x_1 is position, x_2 is velocity, and u is acceleration due to the thrust. We suppose that the thrust-acceleration is not constrained; that is, we impose only

$$|u(t)| < \infty. \quad (13.57)$$

The initial and terminal states as well as the transfer time are prescribed; namely,

$$\begin{aligned} x_1(t_0) - x_1^0 &= 0, \\ x_2(t_0) - x_2^0 &= 0, \\ x_1(t_1) &= 0, \\ x_2(t_1) &= 0, \end{aligned} \tag{13.58}$$

and

$$t_1 - t_0 = \tau_1, \tag{13.59}$$

where x_1^0, x_2^0, and $\tau_1 > 0$ are given.

The quantity to be minimized is the amount of fuel consumed; an equivalent cost is given by

$$\int_{t_0}^{t_1} u^2(t)\, dt. \tag{13.60}$$

This is a problem of the kind discussed in Section 13.7.

In order to deduce candidates for optimal control we employ the necessary conditions of Chapter 11 as modified in Section 13.7. The state equations (13.56) and the cost (13.60) lead to

$$H(\lambda, y, u) = \lambda_0 u^2 + \lambda_1 x_2 + \lambda_2 u \tag{13.61}$$

so that the adjoint equations are

$$\begin{aligned} \dot{\lambda}_0(t) &= 0, \\ \dot{\lambda}_1(t) &= 0, \\ \dot{\lambda}_2(t) &= -\lambda_1(t) \end{aligned}$$

whence

$$\begin{aligned} \lambda_0(t) &\equiv \text{constant}, \\ \lambda_1(t) &= \lambda_1(t_1), \\ \lambda_2(t) &= \lambda_2(t_1) + \lambda_1(t_1)(t_1 - t). \end{aligned} \tag{13.62}$$

Chap. 13 • Some Generalizations

Of course, in view of the condition (c) of Theorem 11.1, $\lambda_0(t) \equiv \text{constant} \leq 0$.

In deducing an extremal control, $u(\cdot):[t_0, t_1] \to R^1$, we consider two possibilities:

(i) $\lambda_0(t) \equiv 0$. In this event, one cannot have

$$\lambda_1(t_1) = \lambda_2(t_1) = 0 \tag{13.63}$$

since (13.62) and (13.63), together with $\lambda_0(t) = 0$, would result in the trivial solution of the adjoint equations, whereas the maximum principle, Theorem 11.1, requires the existence of a nonzero solution. Thus it is not possible that $\lambda_2(t) \equiv 0$ on a nonzero interval, and hence, in view of (13.57), the condition (a) of Theorem 11.1 cannot be met.

(ii) $\lambda_0(t) \equiv \text{constant} \stackrel{\triangle}{=} \lambda_0 < 0$. Since the coefficient of u^2 in (13.61) is negative, $u(t)$ maximizes that expression only if

$$2\lambda_0 u(t) + \lambda_2(t) = 0 \tag{13.64}$$

whence

$$u(t) = c_1 + c_2 \tau \tag{13.65}$$

where, as a consequence of (13.62),

$$c_1 \stackrel{\triangle}{=} -\frac{\lambda_2(t_1)}{2\lambda_0},$$

$$c_2 \stackrel{\triangle}{=} -\frac{\lambda_1(t_1)}{2\lambda_0},$$

$$\tau \stackrel{\triangle}{=} t_1 - t.$$

On integrating the state equations with control (13.65) we obtain

$$c_1 = \frac{6}{\tau_1^2}\left(x_1^0 + \frac{1}{3}x_2^0 \tau_1\right),$$

$$c_2 = -\frac{12}{\tau_1^3}\left(x_1^0 + \frac{1}{2}x_2^0 \tau_1\right) \tag{13.66}$$

so that, with the *time-to-go* τ as the independent variable, $\bar{u}(\cdot):[0,\tau_1]\to R^1$ given by

$$\bar{u}(\tau) = \frac{2}{\tau_1^2}\left[(3x_1^0 + x_2^0\tau_1) - (6x_1^0 + 3x_2^0\tau_1)\frac{\tau}{\tau_1}\right] \quad (13.67)$$

is an extremal control. The cost corresponding to the control (13.67) is

$$\int_0^{\tau_1} \bar{u}^2(\tau)\,d\tau = \frac{4}{\tau_1^3}\left[3(x_1^0)^2 + 3x_1^0 x_2^0 \tau_1 + (x_2^0\tau_1)^2\right]. \quad (13.68)$$

The extremal control given by (13.67) is unique. Subsequently, upon employing sufficient conditions, it can be shown that it is indeed optimal (Exercise 15.6).

13.9. The Simplest Problem of the Calculus of Variations

In Part I we discussed the so-called simplest problem with fixed end points, namely, that of minimizing the value of a given integral

$$\int_{t_0}^{t_1} f[t, x(t), \dot{x}(t)]\,dt$$

with respect to all piecewise smooth functions $x(\cdot):[t_0, t_1]\to R^1$ with prescribed interval $[t_0, t_1]$ and end values $x(t_0)$ and $x(t_1)$. We can pose this problem as one of optimal control by letting

$$x_1 \triangleq x,$$

$$u \triangleq r,$$

$$\hat{f}_0(x_1, t, u) \triangleq f(t, x_1, u),$$

and considering the state equation

$$\dot{x}_1(t) = u(t) \quad (13.69)$$

Chap. 13 • Some Generalizations

with the control constraint set $U = R^1$.

The initial and terminal values of $x_1(\cdot)$ and of t are prescribed; that is, we have

$$\hat{\theta}_1^0(x_1, t) = x_1 - x_1^0 = 0,$$

$$\hat{\theta}_2^0(x_1, t) = t - t_0 = 0,$$

and

$$\hat{\theta}_1^1(x_1, t) = x_1 - x_1^1 = 0,$$

$$\hat{\theta}_2^1(x_1, t) = t - t_1 = 0.$$

The cost for the control problem is

$$\int_{t_0}^{t_1} \hat{f}_0[x_1(t), t, u(t)] \, dt. \tag{13.70}$$

We are now ready to utilize the results of Section 13.6. In view of (13.69) and (13.70) we have

$$\hat{H}(\lambda, y, t, u) = \lambda_0 \hat{f}_0(x_1, t, u) + \lambda_1 u$$

so that the adjoint equations (13.48) become

$$\dot{\lambda}_0(t) = 0,$$

$$\dot{\lambda}_1(t) = \lambda_0(t) \frac{\partial \hat{f}_0[x_1^*(t), t, u^*(t)]}{\partial x_1}. \tag{13.71}$$

Since $|u^*(t)| < \infty$, condition (13.45) implies a local maximum so that it is necessary that

$$\frac{\partial \hat{H}[\lambda(t), y^*(t), t, u^*(t)]}{\partial u} = 0$$

and

$$\frac{\partial^2 \hat{H}[\lambda(t), y^*(t), t, u^*(t)]}{\partial u^2} \leq 0,$$

namely,

$$\lambda_0(t)\frac{\partial \hat{f}_0[x_1^*(t), t, u^*(t)]}{\partial u} + \lambda_1(t) = 0 \qquad (13.72)$$

and

$$\lambda_0(t)\frac{\partial^2 \hat{f}_0[x_1^*(t), t, u^*(t)]}{\partial u^2} \leq 0. \qquad (13.73)$$

Since we require a nonzero solution of (13.71), $\lambda_0(t) \neq 0$; hence the condition (c) of Theorem 11.1 implies that

$$\lambda_0(t) \equiv \text{constant} \stackrel{\triangle}{=} \lambda_0 < 0. \qquad (13.74)$$

Since $\lambda_1(\cdot)$ is differentiable for all $t \in [t_0, t_1]$ with the possible exception of a finite number of points of discontinuity of $u^*(\cdot)$ where it has a right and left derivative, we can differentiate (13.72) on each subinterval of continuity of $u^*(\cdot)$; upon using the second of (13.71) together with (13.74) we obtain

$$\frac{d}{dt}\frac{\partial \hat{f}_0[x_1^*(t), t, u^*(t)]}{\partial u} - \frac{\partial \hat{f}_0[x_1^*(t), t, u^*(t)]}{\partial x_1} = 0.$$

On reverting to the original notation we recognize this equation as the Euler–Lagrange equation (2.22).

In view of (13.72) and (13.74) we have

$$\hat{H}[\lambda(t), y^*(t), t, u] = \lambda_0 \left\{ \hat{f}_0[x_1^*(t), t, u] - u\frac{\partial f_0[x_1^*(t), t, u^*(t)]}{\partial u} \right\} \qquad (13.75)$$

so that condition (13.45) becomes

$$\hat{f}_0[x_1^*(t), t, u] - \hat{f}_0[x_1^*(t), t, u^*(t)] - [u - u^*(t)]\frac{\partial f_0[x_1^*(t), t, u^*(t)]}{\partial u} \geq 0$$

for all $u \in U = R^1$. In terms of the original notation this condition is the

Weierstrass condition (5.2). On the other hand, the Legendre condition (5.10) follows at once from (13.73) with (13.74).

Because of (13.74) we can write (13.72) as

$$\frac{\partial \hat{f}_0[x_1^*(t), t, u^*(t)]}{\partial u} = -\frac{\lambda_1(t)}{\lambda_0}.$$

As a consequence of the continuity of $\lambda_1(\cdot)$, this relation implies the corner condition (7.1).

Finally we invoke condition (13.47). For the problem at hand it is

$$\lambda_0(t)\hat{f}_0[x_1^*(t), t, u^*(t)] + \lambda_1(t)u^*(t)$$

$$= \text{constant} + \int_{t_0}^{t} \frac{\partial \hat{H}[\lambda(\tau), y^*(\tau), \tau, u^*(\tau)]}{\partial \tau} d\tau,$$

or, with (13.72) and (13.74),

$$\lambda_0 \left\{ \hat{f}_0[x_1^*(t), t, u^*(t)] - u^*(t) \frac{\partial \hat{f}_0[x_1^*(t), t, u^*(t)]}{\partial u} \right\}$$

$$= \text{constant} + \int_{t_0}^{t} \frac{\partial \hat{H}[\lambda(\tau), y^*(\tau), \tau, u^*(\tau)]}{\partial \tau} d\tau.$$

In terms of the original notation this relation implies the corner condition (7.2).

Thus we have recovered all but one—the Jacobi condition—of the necessary conditions derived in Part I.

13.10. State-Dependent Control Constraints

Thus far we have admitted controls whose values belong to a prescribed subset, U, of the control space, R^m. Now we consider control constraints which depend on the state, x.

Let

$$\hat{U}(\cdot): R^n \to \text{all nonempty subsets of } R^m \qquad (13.76)$$

be a given *set-valued* function; that is, given $x \in R^n$ there is a *set* $\hat{U}(x) \subset R^m$. Since the set of control values depends on the state we must modify the definition of an admissible control. Now a control $u(\cdot): [t_0, t_1] \to R^m$ that generates the solution $x(\cdot): [t_0, t_1] \to R^n$, $x(t_0) = x^0$, is *admissible at* x^0 if and only if it is piecewise continuous and

$$u(t) \in \hat{U}[x(t)] \qquad \forall t \in [t_0, t_1]. \qquad (13.77)$$

This alteration in the definition of admissibility does not affect the definition of *feasibility at* x^0 provided we replace "admissible" by "admissible at x^0."

Hereafter we suppose that the set-valued function $\hat{U}(\cdot)$ is defined by

$$\begin{aligned}\phi_i(x, u) &= 0, & i &= 1, 2, \ldots, l, \\ \psi_i(x, u) &\leq 0, & i &= 1, 2, \ldots, s,\end{aligned} \qquad (13.78)$$

where the $\phi_i(\cdot): R^n \times R^m \to R^1$ and the $\psi_i(\cdot): R^n \times R^m \to R^1$ are prescribed functions of class C^1. In other words, $u(\cdot): [t_0, t_1] \to R^m$ is admissible at x^0 if

$$\begin{aligned}\phi_i[x(t), u(t)] &= 0, & i &= 1, 2, \ldots, l, \\ \psi_i[x(t), u(t)] &\leq 0, & i &= 1, 2, \ldots, s,\end{aligned}$$

for all $t \in [t_0, t_1]$.

We are concerned with the problem of optimal control for systems of the type discussed in Chapter 10 with the generalization of the control constraint set from a constant set, U, to a state-dependent one, $\hat{U}(x)$. In particular we wish to know how the necessary conditions derived in Chapter 11 are affected by this generalization. In order to permit us to use earlier results we place additional restrictions on the functions $f_i(\cdot): R^n \times R^m \to R^1$, $i = 0, 1, 2, \ldots, n$. In this section we assume that these functions are of class C^1.

We must now retrace our steps to see where and how our earlier results are affected. In doing so we find that the discussion of the transfer of the tangent plane $T_{\Sigma(C)}(y^*(t_0))$ in Section 11.4 is predicated on the constancy of the control constraint set, U. Recall that the solution (11.20), used in the proof of Lemma 11.2, is generated by the control $u^*(\cdot):[t_0, t_1^*] \to R^m$ which is optimal at x^0. In the case of a constant control constraint set, U, the admissibility of $u^*(\cdot)$ is not affected by a change of the initial state from $y^*(t_0)$ to $y^*(t_0) + \varepsilon \eta^0 + o(\varepsilon)$. However, if the control constraint set is state dependent, one cannot be certain that

$$u^*(t) \in \hat{U}(x(t)) \tag{13.79}$$

where

$$y(\cdot) = \left[\begin{array}{c} x_0(\cdot) \\ \hline x(\cdot) \end{array}\right] : [t_0, t_1^*] \to R^{n+1}, \ y(t_0) \neq y^*(t_0),$$

is a solution generated by $u^*(\cdot)$. Indeed we must alter the arguments employed in Section 11.4 in order to assure the satisfaction of (13.79).

Let $[t_0, t_1^*)$ be the union of half-open intervals and let $[t_\alpha, t_\beta)$ denote such a subinterval. Suppose that the same $r \leq s$ inequality constraints are limiting at all $t \in [t_\alpha, t_\beta)$; that is, renumbering if necessary,

$$\psi_i[x^*(t), u^*(t)] = 0, \quad i = 1, 2, \ldots, r, \tag{13.80}$$

and

$$\psi_i[x^*(t), u^*(t)] < 0, \quad i = r+1, r+2, \ldots, s. \tag{13.81}$$

Now consider

$$\begin{aligned} \phi_i(x, u) &= 0, \quad i = 1, 2, \ldots, l, \\ \psi_i(x, u) &= 0, \quad i = 1, 2, \ldots, r, \end{aligned} \tag{13.82}$$

at $x = x^*(t)$, $t \in [t_\alpha, t_\beta)$, and $u = \hat{u} \in \hat{U}(x^*(t))$. We assume that $l + r \leq m$ and let

$$u^c \triangleq [u_1 \ u_2 \cdots u_{l+r}]^T,$$

$$u^a \triangleq [u_{l+r+1} \ u_{l+r+2} \cdots u_m]^T$$

so that

$$u = \left[\begin{array}{c} u^c \\ \hline u^a \end{array}\right].$$

Also define the function

$$R(\cdot): R^{n+1} \times R^{m-l-r} \times R^{l+r} \to R^{l+r}$$

by

$$\begin{aligned} R_i(y, u^a, u^c) &= \phi_j(x, u), & i=j=1,2,\ldots,l, \\ R_i(y, u^a, u^c) &= \psi_j(x, u), & i=l+j, & j=1,2,\ldots,r. \end{aligned} \qquad (13.83)$$

Note that $R(\cdot)$ is constant with respect to x_0. In view of (13.83) one can state (13.82) as

$$R[y^*(t), \hat{u}^a, \hat{u}^c] = 0. \qquad (13.84)$$

We assume that the matrix

$$\begin{bmatrix} \dfrac{\partial R_1(y, u^a, u^c)}{\partial u_1} & \cdots & \dfrac{\partial R_1(y, u^a, u^c)}{\partial u_m} \\ \vdots & & \vdots \\ \dfrac{\partial R_{l+r}(y, u^a, u^c)}{\partial u_1} & \cdots & \dfrac{\partial R_{l+r}(y, u^a, u^c)}{\partial u_m} \end{bmatrix}$$

has maximum rank, $l+r$, at $y=y^*(t)$, $t \in [t_\alpha, t_\beta)$, and all $u = \hat{u} \in U(x^*(t))$. Then it follows from the implicit function theorem (for instance, see Refs. 13.4 and 13.5) that there are a neighborhood, $N(\delta, z)$, of the point

$$z = \begin{bmatrix} y^*(t) \\ --- \\ \hat{u}^a \end{bmatrix} \in R^{n+1} \times R^{m-l-r}$$

and a function $u^c(\cdot): N(\delta, z) \to R^{l+r}$ of class C^1 such that

$$R[y, u^a, u^c(y, u^a)] = 0 \qquad (13.85)$$

for all

$$\begin{bmatrix} y \\ -- \\ u^a \end{bmatrix} \in N(\delta, z),$$

Chap. 13 • Some Generalizations

and

$$\hat{u}^c = u^c\bigl[y^*(t), \hat{u}^a\bigr].$$

Now consider the function $u^a(\cdot):[t_\alpha, t_\beta) \to R^{m-l-r}$ such that

$$u_i^a(t) = u_{l+r+i}^*(t), \qquad i=1,2,\ldots, m-l-r.$$

Then consider the trajectory which emanates from $y^*(t_0) + \varepsilon \eta^0 + o(\varepsilon) \in \Sigma(C)$ and is generated by the control $u(\cdot):[t_0, t_1^*] \to R^m$ where, for $t \in [t_\alpha, t_\beta)$,

$$u(t) = \begin{bmatrix} u^c(y(t), u^a(t)) \\ ----- \\ u^a(t) \end{bmatrix}. \tag{13.86}$$

As a consequence of the dependence of a solution on the initial conditions (for instance, see Ref. 13.6), the solution $y(\cdot):[t_0, t_1^*] \to R^{n+1}$, $y(t_0) = y^*(t_0) + \varepsilon \eta^0 + o(\varepsilon)$, generated by $u(\cdot)$ is such that

$$y(t) = y^*(t) + \varepsilon \eta(t) + o(t, \varepsilon), \tag{13.87}$$

with (11.13) replaced by the variational equation

$$\dot{\eta}(t) = \left[\frac{\partial h}{\partial y}\right]\eta(t) + \left[\frac{\partial h}{\partial u^c}\right]\kappa(t) \tag{13.88}$$

for $t \in [t_\alpha, t_\beta)$, where

$$\frac{\partial h}{\partial y} \triangleq \begin{bmatrix} 0 & \dfrac{\partial f_0(x^*(t), u^*(t))}{\partial x_1} & \cdots & \dfrac{\partial f_0(x^*(t), u^*(t))}{\partial x_n} \\ \vdots & \vdots & & \vdots \\ 0 & \dfrac{\partial f_n(x^*(t), u^*(t))}{\partial x_1} & \cdots & \dfrac{\partial f_n(x^*(t), u^*(t))}{\partial x_n} \end{bmatrix},$$

$$\frac{\partial h}{\partial u^c} \triangleq \begin{bmatrix} \dfrac{\partial f_0(x^*(t), u^*(t))}{\partial u_1} & \cdots & \dfrac{\partial f_0(x^*(t), u^*(t))}{\partial u_{l+r}} \\ \vdots & & \vdots \\ \dfrac{\partial f_n(x^*(t), u^*(t))}{\partial u_1} & \cdots & \dfrac{\partial f_n(x^*(t), u^*(t))}{\partial u_{l+r}} \end{bmatrix},$$

and $\kappa(\cdot):[t_\alpha, t_\beta] \to R^{l+r}$ is such that

$$\kappa_i(t) = \sum_{j=0}^n \frac{\partial u_i^c[y^*(t), u^a(t)]}{\partial x_j} \eta_j(t).$$

Let ε be sufficiently small so that

$$\begin{bmatrix} y(t) \\ ---- \\ u^a(t) \end{bmatrix} \in N(\delta, z) \quad \text{for } z = \begin{bmatrix} y^*(t) \\ ---- \\ u^a(t) \end{bmatrix}, \quad t \in [t_\alpha, t_\beta].$$

Then it follows from (13.85), namely,

$$R[y(t), u^a(t), u^c(y(t), u^a(t))] = 0, \tag{13.89}$$

that the constraints (13.82) are satisfied at $x = x(t)$ and $u = u(t)$. Thus, to prove that $u(\cdot)$ is admissible, we need only show that the remaining constraints are met; that is,

$$\psi_i[x(t), u(t)] < 0, \quad i = r+1, r+2, \ldots, s. \tag{13.90}$$

Since $\psi_i[x(t), u(t)]$ differs from $\psi_i[x^*(t), u^*(t)]$ by terms of order ε, it follows from (13.81) that (13.90) is satisfied for sufficiently small ε.

Furthermore, by expanding (13.89) using (13.87), one obtains

$$\left[\frac{\partial R}{\partial y}\right] \eta(t) + \left[\frac{\partial R}{\partial u^c}\right] \kappa(t) = 0, \tag{13.91}$$

where

$$\frac{\partial R}{\partial y} \triangleq \begin{bmatrix} 0 & \frac{\partial R_1}{\partial x_1} & \cdots & \frac{\partial R_1}{\partial x_n} \\ \vdots & \vdots & & \vdots \\ 0 & \frac{\partial R_{l+r}}{\partial x_1} & \cdots & \frac{\partial R_{l+r}}{\partial x_n} \end{bmatrix},$$

$$\frac{\partial R}{\partial u^c} \triangleq \begin{bmatrix} \frac{\partial R_1}{\partial u_1} & \cdots & \frac{\partial R_1}{\partial u_{l+r}} \\ \vdots & & \vdots \\ \frac{\partial R_{l+r}}{\partial u_1} & \cdots & \frac{\partial R_{l+r}}{\partial u_{l+r}} \end{bmatrix}$$

Chap. 13 • Some Generalizations   173

and the arguments of $\partial R_i/\partial x_j$ and $\partial R_i/\partial u_j$, respectively, are $y^*(t)$, $u^a(t)$ and $u^c[y^*(t), u^a(t)]$.

Now, on solving (13.91) for $\kappa(t)$ and substitution in (13.88), we obtain

$$\dot{\eta}(t) = \left[\frac{\partial h}{\partial y} - \frac{\partial h}{\partial u^c} \left(\frac{\partial R}{\partial u^c} \right)^{-1} \frac{\partial R}{\partial y} \right] \eta(t) \tag{13.92}$$

for $t \in [t_\alpha, t_\beta)$. This equation replaces (11.13). The solutions of (13.92) define a linear transformation $\hat{A}(t, t_\alpha)$ such that

$$\eta(t) = \hat{A}(t, t_\alpha) \eta(t_\alpha)$$

for $t \in [t_\alpha, t_\beta)$.

Finally we note that conditions (11.9)–(11.10) are unaffected since $u^c(\cdot)$ is of class C^1 so that, given $u(\cdot)$ with $u(t) \in \hat{U}(x^*(t))$, there is a corresponding solution $y(\cdot): [t, t+\Delta t] \to R^{n+1}$, $\Delta t > 0$, $y(t) = y^*(t)$.

All subsequent arguments in Section 11.4 remain unchanged provided $\hat{A}(t, t_\alpha)$ is used for the transfer of the tangent plane $T_{\Sigma(C)}(y^*(t_\alpha))$, $t \in [t_\alpha, t_\beta)$. Consequently Theorem 11.1 remains in force with U replaced by $\hat{U}(x^*(t))$ and the adjoint equations replaced by those corresponding to (13.92), namely, by

$$\dot{\lambda}^T(t) = -\lambda^T(t) \frac{\partial h}{\partial y} + \nu^T(t) \frac{\partial R}{\partial y} \tag{13.93}$$

where

$$\nu^T(t) \triangleq \lambda^T(t) \frac{\partial h}{\partial u^c} \left(\frac{\partial R}{\partial u^c} \right)^{-1}. \tag{13.94}$$

It must be noted that the adjoint equations may be different on the subintervals $[t_\alpha, t_\beta)$ of $[t_0, t_1^*)$ since different constraints may be limiting on different intervals. However, $\lambda(\cdot): [t_0, t_1^*] \to R^{n+1}$ is continuous, that is, $\lambda(t_\beta - 0) = \lambda(t_\beta + 0)$, if $\Gamma^*(C)$ is a regular optimal trajectory.

In order to obtain a single adjoint equation for the entire interval $[t_0, t_1^*]$, let us introduce the function $\mu(\cdot): [t_0, t_1^*] \to R^m$ such that, for $t \in [t_\alpha, t_\beta)$,

$$\mu(t) \triangleq [\nu_1(t) \; \nu_2(t) \cdots \nu_{l+r}(t) \; 0 \cdots 0]^T \tag{13.95}$$

with the zero components corresponding to the strict inequalities (13.81), that is, to the nonlimiting constraints. Since these may differ on different subintervals of $[t_0, t_1^*)$, the function $\mu(\cdot)$ may be discontinuous at the end points of such subintervals so that $\mu(t_\beta - 0) \neq \mu(t_\beta + 0)$, as well as at points of discontinuity of $u^*(\cdot)$; thus, $\mu(\cdot)$ is piecewise continuous.

Also let $S(\cdot): R^{n+1} \times R^m \to R^{l+s}$ be such that

$$\begin{aligned} S_i(y, u) &= \phi_j(x, u), & i = j = 1, 2, \ldots, l, \\ S_i(y, u) &= \psi_j(x, u), & i = l + j, & j = 1, 2, \ldots, s. \end{aligned} \qquad (13.96)$$

Finally define

$$\frac{\partial S}{\partial y} \triangleq \begin{bmatrix} 0 & \dfrac{\partial S_1(y^*(t), u^*(t))}{\partial x_1} & \cdots & \dfrac{\partial S_1(y^*(t), u^*(t))}{\partial x_n} \\ \vdots & \vdots & & \vdots \\ 0 & \dfrac{\partial S_{l+s}(y^*(t), u^*(t))}{\partial x_1} & \cdots & \dfrac{\partial S_{l+s}(y^*(t), u^*(t))}{\partial x_n} \end{bmatrix}.$$

Then we obtain the single adjoint equation

$$\dot{\lambda}^T(t) = -\lambda^T(t) \frac{\partial h}{\partial y} + \mu^T(t) \frac{\partial S}{\partial y} \qquad (13.97)$$

for $t \in [t_0, t_1^*]$.

13.11. A Time-Optimal Regulator with Velocity-Dependent Control Bounds

Consider the system with the state equations

$$\dot{x}_1(t) = x_2(t), \qquad \dot{x}_2(t) = u(t) \qquad (13.98)$$

and with the control constraint set, $\hat{U}(x)$, defined by

$$|u| \leq x_2^2,$$

or, corresponding to (13.78), by

$$\psi_1(x, u) = -(u + x_2^2) \leq 0, \qquad \psi_2(x, u) = u - x_2^2 \leq 0. \qquad (13.99)$$

Chap. 13 • Some Generalizations

The end states are prescribed; that is,

$$x_1(t_0) = x_1^0, \qquad x_2(t_0) = x_2^0,$$

and

$$x_1(t_1) = x_1^1, \qquad x_2(t_1) = x_2^1.$$

The cost to be minimized is

$$t_1 - t_0 = \int_{t_0}^{t_1} dt \tag{13.100}$$

so that, if t stands for time, the problem is that of the time-optimal transfer between given end states.

Before employing the maximum principle, as modified in Section 13.10, to deduce an extremal control, a few words are in order concerning the region, E, of the initial states for which there are feasible controls, that is, from which the given terminal state $x^1 = [x_1^1 \ x_2^1]^T$ can be reached by means of an admissible control. As a consequence of (13.98) and (13.99) we have

$$\left| \frac{dx_2(t)}{dx_1(t)} \right| \leq |x_2(t)|.$$

For instance, if $x_2(t) > 0$, then

$$-x_2(t) \leq \frac{dx_2(t)}{dx_1(t)} \leq x_2(t),$$

so that the region E is bounded by the exponential curves

$$x_2 = k_+ e^{x_1}, \qquad k_+ \stackrel{\triangle}{=} x_2^1 e^{-x_1^1},$$

$$x_2 = k_- e^{-x_1}, \qquad k_- \stackrel{\triangle}{=} x_2^1 e^{x_1^1}.$$

This situation, as well as the analogous one for $x_2(t) < 0$, are shown in Figure 13.4. Note that the boundary belongs to the region E; that is, E is closed.

The direction of increasing time along a trajectory in R^2 is determined by the first of the state equations (13.98); for instance, if $x_2^0 > 0$, then

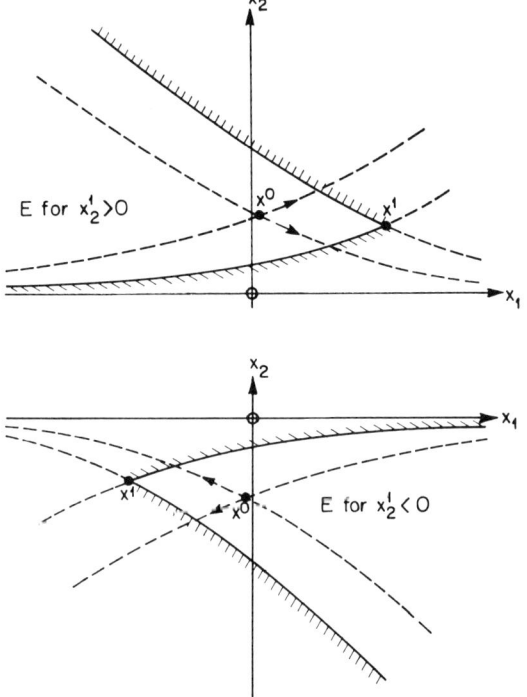

Figure 13.4. Initial state regions.

$x_2(\tau) > 0$ for $\tau \in [t_0, t]$ so that

$$t > t_0 \Rightarrow \int_{t_0}^{t} x_2(\tau)\, d\tau > 0 \Rightarrow x_1(t) > x_1^0.$$

Now let us turn to the determination of an extremal control. In view of (13.98) and (13.100) we have

$$H(\lambda, y, u) = \lambda_0 + \lambda_1 x_2 + \lambda_2 u. \tag{13.101}$$

To determine the appropriate adjoint equations we must consider three possibilities; namely, if $u(\cdot): [t_0, t_1] \to R^1$ is an extremal control with the corresponding solution $x(\cdot): [t_0, t_1] \to R^2$, then it follows from (13.99) that

(i) $|u(t)| < x_2^2(t)$

or

(ii) $u(t) + x_2^2(t) = 0$

or

(iii) $u(t) - x_2^2(t) = 0$.

Now, according to (13.95), we have

$$\mu(t) = \begin{cases} [0 \ 0]^T & \text{for (i)}, \\ [\nu_1(t) \ 0]^T & \text{for (ii)}, \\ [0 \ \nu_2(t)]^T & \text{for (iii)}, \end{cases}$$

where, by (13.94),

$$\nu_1(t) = -\lambda_2(t), \qquad \nu_2(t) = \lambda_2(t).$$

Consequently, (13.97) becomes

$$\dot{\lambda}_0(t) = 0,$$
$$\dot{\lambda}_1(t) = 0,$$
$$\dot{\lambda}_2(t) = \begin{cases} -\lambda_1(t) & \text{for (i)}, \\ -\lambda_1(t) + 2\lambda_2(t) x_2(t) & \text{for (ii)}, \\ -\lambda_1(t) - 2\lambda_2(t) x_2(t) & \text{for (iii)}. \end{cases} \qquad (13.102)$$

As a result of (13.101) and the condition (b) of Theorem 11.1 we have

$$\lambda_0(t) + \lambda_1(t) x_2(t) + \lambda_2(t) u(t) = 0 \qquad (13.103)$$

so that, in view of (13.102),

$$\lambda_2(t) \equiv 0 \Rightarrow \dot{\lambda}_2(t) \equiv 0 \Rightarrow \lambda_1(t) \equiv 0 \Rightarrow \lambda_0(t) \equiv 0.$$

Since we require a nonzero solution of the adjoint equations (13.102), $\lambda_2(\cdot)$ cannot vanish on a nonzero subinterval of $[t_0, t_1]$. Thus it follows from the modified condition (a) of Theorem 11.1 that an extremal control is always limiting; that is,

$$\begin{aligned} u(t) &= x_2^2(t) & \text{if } \lambda_2(t) > 0, \\ u(t) &= -x_2^2(t) & \text{if } \lambda_2(t) < 0. \end{aligned} \qquad (13.104)$$

The solutions generated by the controls for which

$$u(t) \equiv x_2^2(t) \quad \text{or} \quad u(t) \equiv -x_2^2(t)$$

are obtained by integrating (13.98). For the former control this leads to

$$x_2(t) = c_+ e^{x_1(t)}, \qquad c_+ = \text{constant},$$

while for the latter it results in

$$x_2(t) = c_- e^{-x_1(t)}, \qquad c_- = \text{constant}.$$

These solutions bear out our earlier contention concerning the region E. In particular, the terminal portion of a trajectory generated by an extremal control—that is, the portion that contains the terminal state x^1—belongs to the boundary of E. Furthermore, for $u(t) \equiv x_2^2(t)$,

$$k_+ > 0 \quad (x_2^1 > 0) \Rightarrow \begin{cases} \dot{x}_1(t) = x_2(t) > 0, \\ \dot{x}_2(t) = x_2^2(t) > 0, \end{cases}$$

and

$$k_+ < 0 \quad (x_2^1 < 0) \Rightarrow \begin{cases} \dot{x}_1(t) = x_2(t) < 0, \\ \dot{x}_2(t) = x_2^2(t) > 0. \end{cases}$$

Similarly, for $u(t) \equiv -x_2^2(t)$,

$$k_- > 0 \quad (x_2^1 > 0) \Rightarrow \begin{cases} \dot{x}_1(t) = x_2(t) > 0, \\ \dot{x}_2(t) = -x_2^2(t) < 0, \end{cases}$$

$$k_- < 0 \quad (x_2^1 < 0) \Rightarrow \begin{cases} \dot{x}_1(t) = x_2(t) < 0, \\ \dot{x}_2(t) = -x_2^2(t) < 0. \end{cases}$$

The next question pertains to the switching sequence for an extremal control. At an instant of switching, say at $t = t_s$, the continuity of $\lambda_2(\cdot)$ implies that $\lambda_2(t_s) = 0$; hence, in accordance with (13.102),

$$\dot{\lambda}_2(t_s) = -\lambda_1(t_s).$$

However,

$$\lambda_1(t) \equiv \text{constant} \stackrel{\triangle}{=} \lambda_1$$

so that there can be at most one zero of $\lambda_2(\cdot)$, and hence at most one switch of $u(\cdot)$. In particular,

$$\lambda_1 > 0 \Rightarrow \dot{\lambda}_2(t_s) < 0 \Rightarrow \begin{cases} \lambda_2(t) > 0, & t < t_s, \\ \lambda_2(t) < 0, & t > t_s, \end{cases}$$

and so

$$u(t) = x_2^2(t) \quad \text{for } t < t_s,$$
$$u(t) = -x_2^2(t) \quad \text{for } t > t_s.$$

Conversely,

$$\lambda_1 < 0 \Rightarrow \dot{\lambda}_2(t_s) > 0 \Rightarrow \begin{cases} \lambda_2(t) < 0, & t < t_s, \\ \lambda_2(t) > 0, & t > t_s, \end{cases}$$

and so

$$u(t) = -x_2^2(t) \quad \text{for } t < t_s,$$
$$u(t) = x_2^2(t) \quad \text{for } t > t_s.$$

Which of these switching sequences corresponds to an extremal control? To answer this question we consider (13.103) at $t = t_s$; namely,

$$\lambda_0(t_s) + \lambda_1 x_2(t_s) = 0. \tag{13.105}$$

If there is a switch, then (13.105) must hold and $\lambda_0(t) \equiv \text{constant} \neq 0$, lest $\lambda_1(t) \equiv 0 \equiv \lambda_2(t)$. Hence, if there is a switch, $\lambda_0(t) \equiv \text{constant} < 0$ in conformity with the condition (c) of Theorem 11.1; consequently,

$$\lambda_1 > 0 \Leftrightarrow x_2(t_s) > 0,$$
$$\lambda_1 < 0 \Leftrightarrow x_2(t_s) < 0.$$

Furthermore we recall that the sign of $x_2(\cdot)$ is invariant; that is,

$$x_2(t_s) > 0 \Leftrightarrow x_2^1 > 0,$$
$$x_2(t_s) < 0 \Leftrightarrow x_2^1 < 0.$$

Thus we conclude that the switching sequence is

$$u(t) = x_2^2(t) \quad \text{for } t < t_s,$$
$$u(t) = -x_2^2(t) \quad \text{for } t > t_s,$$

if $x_2^1 > 0$, and

$$u(t) = -x_2^2(t) \quad \text{for } t < t_s,$$
$$u(t) = x_2^2(t) \quad \text{for } t > t_s,$$

if $x_2^1 < 0$.

Switching occurs when the trajectory reaches the boundary of the region E since the terminal portion of the trajectory belongs to the boundary of that region; this is illustrated in Figure 13.5. We leave it as an exercise for the reader to derive the switching time, t_s, as a function of the given end states (Exercise 13.10).

If the initial state belongs to the boundary of the region E, no switch takes place.

The value of the cost, that is, of the transfer time, $t_1 - t_0$, corresponding to an extremal control is easily computed by integrating the state equations

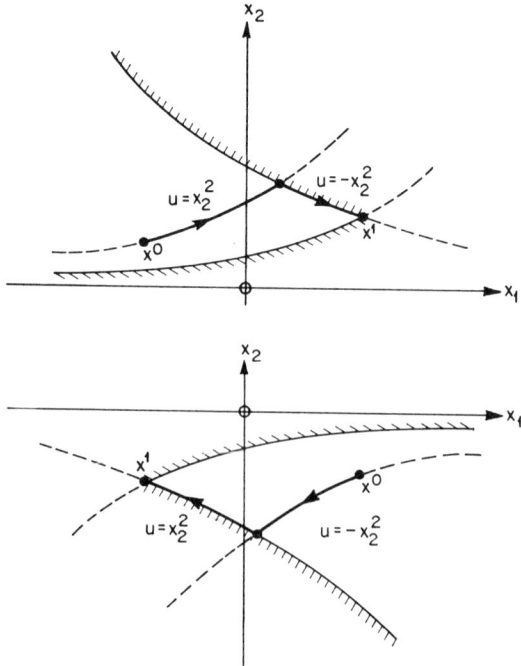

Figure 13.5. Switching sequences.

(13.98) (Exercise 13.10). For instance, if $x_2^1 > 0$, then

$$t_1 - t_0 = \frac{1}{x_2^0} + \frac{1}{x_2^1} - 2\left(\frac{1}{x_2^0 x_2^1}\right)^{1/2} \exp\left(\frac{x_1^0 - x_1^1}{2}\right).$$

Clearly, for given end states, the extremal control is unique; nonetheless we cannot yet assert that it is optimal (Exercise 15.7).

13.12. Isoperimetric Constraints

The problem of optimal control, as posed in Chapter 10 and treated in Chapter 11, is that of determining an admissible control which "steers" the state of the system from a given initial one, x^0 — or more generally from a state in a given initial state set, θ^0 — to a state in a prescribed target set, θ^1, while minimizing the value of an associated cost. In this problem a control $u(\cdot):[t_0,t_1] \to R^m$, generating a solution $x(\cdot):[t_0,t_1] \to R^n$, is admissible provided its value belongs to a given set, U, for all $t \in [t_0,t_1]$. Now we shall introduce additional constraints.

Let $g(\cdot): R^n \times R^m \to R^s$ be a given function of class C^1 and let $l \in R^s$ be a given constant. As before, let (10.1) be the state equations; namely,

$$\dot{x}_j(t) = f_j[x(t), u(t)], \quad j = 1, 2, \ldots, n, \quad (13.106)$$

where the control $u(\cdot):[t_0, t_1] \to R^m$ satisfies

$$u(t) \in U \quad \forall t \in [t_0, t_1] \quad (13.107)$$

and generates the solution $x(\cdot):[t_0, t_1] \to R^n$ such that

$$x(t_0) \in \theta^0, \quad x(t_1) \in \theta^1 \quad (13.108)$$

and such that

$$\int_{t_0}^{t_1} g_i[x(t), u(t)] \, dt = l_i, \quad i = 1, 2, \ldots, \hat{s}, \quad (13.109)$$

$$\int_{t_0}^{t_1} g_i[x(t), u(t)] \, dt \leq l_i, \quad i = \hat{s}+1, \hat{s}+2, \ldots, s. \quad (13.110)$$

As before, the cost is

$$\int_{t_0}^{t_1} f_0[x(t), u(t)]\, dt. \quad (13.111)$$

Thus the problem differs from the one discussed heretofore in that we seek a control subjected to additional constraints which depend on the solution generated by the control. Hence, as in the case of state-dependent control constraints treated in Section 13.10, we modify the definition of an *admissible* control. Now a control $u(\cdot):[t_0, t_1] \to R^m$ that generates the solution $x(\cdot):[t_0, t_1] \to R^n$, $x(t_0) = x^0$, is *admissible at* x^0 if and only if it is piecewise continuous and such that the constraint (13.107) as well as the isoperimetric constraints (13.109)–(13.110) are satisfied. This alteration in the definition of admissibility does not affect the definition of *feasibility at* x^0 provided we replace "admissible" by "admissible at x^0." To deduce necessary conditions for a control that is optimal on θ^0 for the system (13.106)–(13.111) we proceed in a manner similar to that employed in Section 13.4; see also Ref. 13.7.

Let

$$\tilde{x} \triangleq [x_1 \ x_2 \cdots x_{n+2s}]^T,$$

$$\tilde{u} \triangleq [u_1 \ u_2 \cdots u_{m+s}]^T,$$

and consider the augmented system whose state equations are

$$\begin{aligned}
\dot{x}_j(t) &= f_j[x(t), u(t)], & j &= 1, 2, \ldots, n, \\
\dot{x}_{n+i}(t) &= g_i[x(t), u(t)], & i &= 1, 2, \ldots, s, \quad (13.112) \\
\dot{x}_{n+s+i}(t) &= u_{m+i}(t), & i &= 1, 2, \ldots, s,
\end{aligned}$$

where the control $\tilde{u}(\cdot):[t_0, t_1] \to R^{m+s}$ satisfies

$$\tilde{u}(t) \in \tilde{U} \qquad \forall t \in [t_0, t_1] \quad (13.113)$$

with $\tilde{U}$ defined by

$$u \in U,$$

$$u_{m+i} \begin{cases} =0 & \text{for } i=1,2,\ldots, \hat{s}, \\ \leq 0 & \text{for } i=\hat{s}+1, \hat{s}+2, \ldots, s, \end{cases}$$

and generates the solution $\tilde{x}(\cdot):[t_0, t_1] \to R^{n+2s}$ such that $\tilde{x}(t_0) \in \tilde{\theta}^0$, namely,

$$x(t_0) \in \theta^0,$$
$$x_{n+i}(t_0) = 0, \quad i = 1, 2, \ldots, 2s, \quad (13.114)$$

and $\tilde{x}(t_1) \in \tilde{\theta}^1$, namely,

$$x(t_1) \in \theta^1,$$
$$x_{n+i}(t_1) - l_i - x_{n+s+i}(t_1) = 0, \quad i = 1, 2, \ldots, s. \quad (13.115)$$

The cost functional is unchanged; that is, it is

$$\int_{t_0}^{t_1} f_0[x(t), u(t)] \, dt. \quad (13.116)$$

If $u(\cdot):[t_0, t_1] \to R^m$ generates the solution $x(\cdot):[t_0, t_1] \to R^n$ satisfying (13.106)–(13.110), then the control $\tilde{u}(\cdot) = [u_1(\cdot) \; u_2(\cdot) \; \cdots \; u_{m+s}(\cdot)]^T$ generates the solution $\tilde{x}(\cdot) = [x_1(\cdot) \; x_2(\cdot) \; \cdots \; x_{n+2s}(\cdot)]^T$ satisfying (13.112)–(13.115), provided the piecewise continuous functions $u_{m+i}(\cdot)$, $i = 1, 2, \ldots, s$, are such that

$$u_{m+i}(t) = 0, \quad i = 1, 2, \ldots, \hat{s},$$
$$u_{m+i}(t) \leq 0, \quad i = \hat{s}+1, \hat{s}+2, \ldots, s, \quad (13.117)$$
$$\int_{t_0}^{t_1} u_{m+i}(t) \, dt = \int_{t_0}^{t_1} g_i[x(t), u(t)] \, dt - l_i, \quad i = 1, 2, \ldots, s.$$

Furthermore then the values of the costs (13.111) and (13.116) are equal. Thus, if $u^*(\cdot):[t_0, t_1^*] \to R^m$ generates the solution $x^*(\cdot):[t_0, t_1^*] \to R^n$ and is optimal on θ^0 for the system (13.106)–(13.111), then the control $\tilde{u}^*(\cdot) = [u_1^*(\cdot) \; u_2^*(\cdot) \; \cdots \; u_{m+s}^*(\cdot)]^T$ that generates the solution $\tilde{x}^*(\cdot) = [x_1^*(\cdot) \; x_2^*(\cdot) \; \cdots \; x_{n+2s}^*(\cdot)]^T$ is optimal on $\tilde{\theta}^0$ for the system (13.112)–(13.116), provided the $u_{m+i}^*(\cdot)$ satisfy (13.117). Suppose that

$$\int_{t_0}^{t_1^*} g_i[x^*(t), u^*(t)] \, dt - l_i = 0, \quad i = 1, 2, \ldots, r \geq \hat{s},$$

and

$$\int_{t_0}^{t_1^*} g_i[x^*(t), u^*(t)] \, dt - l_i < 0, \quad i = r+1, r+2, \ldots, s.$$

Then one can satisfy (13.117) by choosing

$$u^*_{m+i}(t) \equiv 0, \qquad i=1,2,\ldots,r,$$
$$u^*_{m+i}(t) \equiv \frac{x^*_{n+i}(t^*_1) - l_i}{t^*_1 - t_0}, \qquad i=r+1, r+2,\ldots,s. \qquad (13.118)$$

Thus the *necessary* conditions for the optimality of $\tilde{u}^*(\cdot)$ for the system (13.112)–(13.116) are also necessary for the optimality of $u^*(\cdot)$ for the system (13.106)–(13.111). However, the former is a system of the type treated in Chapter 11 and Section 13.2; consequently Theorem 11.1 and the initial transversality condition (13.8)–(13.9) apply.

Let

$$\tilde{y} \triangleq [x_0 \; x_1 \cdots x_{n+2s}]^T,$$

$$\tilde{\lambda} \triangleq [\lambda_0 \; \lambda_1 \cdots \lambda_{n+2s}]^T.$$

Then, in view of (13.112), we define

$$\tilde{H}(\tilde{\lambda}, \tilde{y}, \tilde{u}) \triangleq \sum_{i=0}^{n} \lambda_i f_i(x, u) + \sum_{i=1}^{s} \lambda_{n+i} g_i(x, u) + \sum_{i=1}^{s} \lambda_{n+s+i} u_{m+i}.$$
$$(13.119)$$

Consequently the adjoint equations for the augmented system are

$$\dot{\lambda}_0(t) = 0,$$
$$\dot{\lambda}_j(t) = -\frac{\partial \tilde{H}[\tilde{\lambda}(t), \tilde{y}^*(t), \tilde{u}^*(t)]}{\partial x_j}, \qquad j=1,2,\ldots,n, \quad (13.120)$$
$$\dot{\lambda}_j(t) = 0, \qquad j=n+1, n+2,\ldots, n+2s.$$

As a consequence of the initial conditions (13.114), the initial transversality condition (13.8)–(13.9) reduces to

$$\sum_{i=1}^{n} \lambda_i(t_0) \eta_i^0 = 0 \qquad (13.121)$$

subject to

$$\sum_{i=1}^{n} \frac{\partial \theta_j^0(x^*(t_0))}{\partial x_i} \eta_i^0 = 0, \qquad j=1,2,\ldots,p. \qquad (13.122)$$

Furthermore, in view of the terminal conditions (13.115), the terminal transversality condition (11.38)–(11.39) leads to

$$\sum_{i=1}^{n} \lambda_i(t_1^*) \eta_i^1 = 0 \qquad (13.123)$$

subject to

$$\sum_{i=1}^{n} \frac{\partial \theta_j^1(x^*(t_1^*))}{\partial x_i} \eta_i^1 = 0, \qquad j=1,2,\ldots,q, \qquad (13.124)$$

and to

$$\lambda_{n+i}(t_1^*) = -\lambda_{n+s+i}(t_1^*), \qquad i=1,2,\ldots,s. \qquad (13.125)$$

The condition (a) of Theorem 11.1, namely,

$$\max_{\tilde{u} \in \tilde{U}} \tilde{H}\left[\tilde{\lambda}(t), \tilde{y}^*(t), \tilde{u}\right] = \tilde{H}\left[\tilde{\lambda}(t), \tilde{y}^*(t), \tilde{u}^*(t)\right]$$

implies that

$$\max_{u \in U} \left\{ \sum_{i=0}^{n} \lambda_i(t) f_i[x^*(t), u] + \sum_{i=1}^{s} \lambda_{n+i}(t) g_i[x^*(t), u] \right\}$$

$$= \sum_{i=0}^{n} \lambda_i(t) f_i[x^*(t), u^*(t)] + \sum_{i=1}^{s} \lambda_{n+i}(t) g_i[x^*(t), u^*(t)], \qquad (13.126)$$

and

$$\lambda_{n+s+i}(t)\left[u_{m+i}^*(t) - u_{m+i}\right] \geq 0 \qquad \forall u_{m+i} \leq 0 \qquad (13.127)$$

for $i = \hat{s}+1, \hat{s}+2, \ldots, s$.

The conditions (13.127), together with (13.118), imply in turn that

$$\lambda_{n+s+i}(t) \geq 0 \quad \text{for } i=\hat{s}+1, \hat{s}+2,\ldots,r,$$
$$\lambda_{n+s+i}(t) \equiv 0 \quad \text{for } i=r+1, r+2,\ldots,s. \tag{13.128}$$

Furthermore, as a consequence of the adjoint equations (13.120) together with (13.125) and (13.128), we have

$$\lambda_{n+i}(t) = -\lambda_{n+s+i}(t) \equiv \text{constant} \overset{\triangle}{=} \beta_i \tag{13.129}$$

for $i=1,2,\ldots,s$, where $\beta_i \leq 0$ for $i=\hat{s}+1, \hat{s}+2,\ldots,s$.

Of course, by the condition (c) of Theorem 11.1,

$$\lambda_0(t) \equiv \text{constant} \leq 0. \tag{13.130}$$

Finally using the condition (b) of Theorem 11.1, namely,

$$\tilde{H}\big[\tilde{\lambda}(t), \tilde{y}^*(t), \tilde{u}^*(t)\big] = 0,$$

together with (13.118), (13.128), and (13.129), we arrive at

$$\sum_{i=0}^{n} \lambda_i(t) f_i[x^*(t), u^*(t)] + \sum_{i=1}^{s} \beta_i g_i[x^*(t), u^*(t)] = 0. \tag{13.131}$$

Before summarizing the conditions which are necessary if $u^*(\cdot)$ generating the solution $x^*(\cdot)$ is optimal on θ^0 for the system (13.106)–(13.111), let us define the function $\overline{H}(\cdot) : R^{n+1} \times R^s \times R^{n+1} \times R^m \to R^1$ by

$$\overline{H}(\lambda, \beta, y, u) \overset{\triangle}{=} H(\lambda, y, u) + \sum_{i=1}^{s} \beta_i g_i(x, u) \tag{13.132}$$

where, as before,

$$H(\lambda, y, u) \overset{\triangle}{=} \sum_{i=0}^{n} \lambda_i f_i(x, u).$$

In terms of $\overline{H}(\cdot)$ the adjoint equations (13.120) lead to

$$\dot{\lambda}_0(t) = 0,$$
$$\dot{\lambda}_j(t) = -\frac{\partial \overline{H}[\lambda(t), \beta, y^*(t), u^*(t)]}{\partial x_j}, \quad j=1,2,\ldots,n. \tag{13.133}$$

Now we have the following theorem pertaining to the system (13.106)–(13.111).

Theorem 13.1. *If the control $u^*(\cdot):[t_0, t_1^*] \to R^m$, generating the solution $x^*(\cdot):[t_0, t_1^*] \to R^n$, is optimal on θ^0 for the system (13.106)–(13.111), then there are a solution $\lambda(\cdot):[t_0, t_1^*] \to R^{n+1}$ of (13.133) and a constant $\beta \in R^s$, where $[\lambda^T(t) \vdots \beta^T] \neq 0$ such that*

(a) $\max\limits_{u \in U} \overline{H}[\lambda(t), \beta, y^*(t), u] = \overline{H}[\lambda(t), \beta, y^*(t), u^*(t)]$,

(b) $\overline{H}[\lambda(t), \beta, y^*(t), u^*(t)] = 0$,

(c) $\lambda_0(t) = \text{constant} \leq 0$

for all $t \in [t_0, t_1^*]$, with (a) and (b) holding for $u^*(\bar{t}-0)$ and $u^*(\bar{t}+0)$ if $u^*(\cdot)$ is discontinuous at $\bar{t} \in (t_0, t_1^*)$,

(d) $\beta_i \leq 0$, $\quad i = \hat{s}+1, \hat{s}+2, \ldots, s, \quad$ *where $\beta_j = 0$ if*

$$\int_{t_0}^{t_1^*} g_j[x^*(t), u^*(t)] \, dt < l_j,$$

and

(e) *the transversality conditions (13.121)–(13.122) and (13.123)–(13.124) are satisfied.*

We conclude this section with the following observations. Recall the optimal control problem with a fixed interval $[t_0, t_1]$, discussed in Section 13.7. There it was shown that the necessary conditions embodied in Theorem 11.1 remain unchanged with the sole exception of the condition (b), which becomes

$$H[\lambda(t), y^*(t), u^*(t)] = \text{constant}.$$

We note now that this problem can be posed as one with the single isoperimetric constraint

$$\int_{t_0}^{t_1} dt = l_1.$$

Then, by (13.132), we have

$$\overline{H}[\lambda(t), \beta, y^*(t), u^*(t)] = H[\lambda(t), y^*(t), u^*(t)] + \beta_1$$

so that the condition (b) of Theorem 13.1 implies that

$$H[\lambda(t), y^*(t), u^*(t)] = -\beta_1 = \text{constant}. \tag{13.134}$$

Thus we recover the result of Section (13.7).

Finally, as in Section 13.4, while it is not required for the purpose of deducing *necessary* conditions, the converse of our claim can be shown to be true; namely, if $\tilde{u}^*(\cdot)$ is optimal on $\tilde{\theta}^0$ for the system (13.112)–(13.116), then $u^*(\cdot)$ is optimal on θ^0 for the original system (13.106)–(13.111).

13.13. Dido's Problem

Recall Dido's Problem, Example 3.5 of Section 3.5, where it was posed as a problem of the calculus of variations. In the guise of an optimal control problem it may be stated as follows. Let t be the arc length, x_1, x_2 be the position coordinates, and u_1, u_2 be the direction cosines of the tangent of a curve in R^2. Then the applicable state equations are

$$\dot{x}_1(t) = u_1(t), \qquad \dot{x}_2(t) = u_2(t), \tag{13.135}$$

where the control is constrained by

$$u_1^2(t) + u_2^2(t) = 1. \tag{13.136}$$

The end conditions are

$$x_1(t_0) = x_2(t_0) = 0, \qquad x_2(t_1) = 0. \tag{13.137}$$

The quantity to be maximized is the area between a curve of given length, l, and the x_1-axis; thus the cost is

$$-\int_{t_0}^{t_1} u_1(t) x_2(t) \, dt \tag{13.138}$$

with the minus sign to account for the minimization of the cost (13.138). The isoperimetric constraint is

$$\int_{t_0}^{t_1} dt = l. \tag{13.139}$$

As pointed out at the end of the last section, one can treat such a problem as one with a fixed interval and apply the results of Section 13.7. We choose to view it as an isoperimetrically constrained one and to employ Theorem 13.1.

Chap. 13 • Some Generalizations

It follows from (13.135), (13.138), and (13.139) that

$$\overline{H}[\lambda, \beta, y, u] = -\lambda_0 u_1 x_2 + \lambda_1 u_1 + \lambda_2 u_2 + \beta$$

so that equations (13.133) lead to

$$\dot{\lambda}_1(t) = 0, \quad \dot{\lambda}_2(t) = \lambda_0(t) u_1(t).$$

The terminal transversality condition (13.123)–(13.124), together with (13.137), implies that

$$\lambda_1(t_1) = 0$$

so that

$$\lambda_1(t) \equiv 0.$$

Now it follows from the condition (a) of Theorem 13.1 that

$$\begin{aligned} u_1(t) &= \frac{-\lambda_0(t) x_2(t)}{\left[\lambda_0^2(t) x_2^2(t) + \lambda_2^2(t)\right]^{1/2}}, \\ u_2(t) &= \frac{\lambda_2(t)}{\left[\lambda_0^2(t) x_2^2(t) + \lambda_2^2(t)\right]^{1/2}}. \end{aligned} \qquad (13.140)$$

It is readily seen that $\lambda_0(t) \neq 0$. If $\lambda_0(t) \equiv \text{constant} = 0$, then $u_1(t) \equiv 0$ so that $\dot{x}_2(t) \equiv 1$; but then the end conditions (13.137) cannot be met. Hence we may set $\lambda_0(t) \equiv -1$. Consequently we have

$$\dot{\lambda}_2(t) = -u_1(t) = -\dot{x}_1(t)$$

whence

$$\lambda_2(t) = -x_1(t) + c_1, \quad c_1 = \text{constant}.$$

Thus, in view of (13.140) with (13.135), we have

$$\frac{u_2(t)}{u_1(t)} = \frac{\dot{x}_2(t)}{\dot{x}_1(t)} = \frac{-x_1(t) + c_1}{x_2(t)}.$$

From this relation one can easily deduce the curve corresponding to the extremal control $[u_1(\cdot) \ u_2(\cdot)]^T$; namely, it is a curve with the slope

$$\frac{dx_2}{dx_1} = -\frac{x_1-c_1}{x_2}$$

so that, in view of (13.137), we obtain

$$(x_1-c_1)^2+x_2^2=c_1^2,$$

which is the equation of a circle of radius c_1 centered at $x_1=c_1$, $x_2=0$. Since we are concerned with the area enclosed by the curve and the x_1-axis, the curve is a semicircle of length l and radius $c_1=l/\pi$.

Since the curve was deduced from necessary conditions we cannot yet assert that it encloses the *maximum* area; see Exercise 15.8.

13.14. Parameter Optimization

Another variant of the optimal control problem, posed in Chapter 10 and treated in subsequent chapters, is one in which the evolution of the state depends not only on the choice of the control *function* $u(\cdot)$ but also on the choice of a *constant* parameter $w \in R^\gamma$. Thus, for prescribed functions

$$\hat{f}_i(\cdot): R^n \times R^\gamma \times R^m \to R^1, \qquad i=0,1,\ldots,n,$$

the state equations are now

$$\dot{x}_i(t)=\hat{f}_i[x(t),w,u(t)], \qquad i=1,2,\ldots,n. \qquad (13.141)$$

We shall suppose that the $\hat{f}_i(\cdot)$, $\partial \hat{f}_i(\cdot)/\partial x_j$ and $\partial \hat{f}_i(\cdot)/\partial w_j$ are continuous on $R^n \times R^\gamma \times R^m$.

The rest of the problem statement is unchanged. The control $u(\cdot)$ and parameter w must be such that $u(\cdot)$ is admissible and the state is transferred between prescribed smooth manifolds; namely, we require that

$$x(t_0) \in \theta^0, \qquad x(t_1) \in \theta^1. \qquad (13.142)$$

The cost associated with such a transfer is

$$\int_{t_0}^{t_1} \hat{f}_0[x(t),w,u(t)] \, dt. \qquad (13.143)$$

Chap. 13 • Some Generalizations

Now we seek an optimal control on θ^0, $u^*(\cdot):[t_0, t_1^*] \to R^m$, as well as an optimal parameter, $w^* \in R^\gamma$, generating the solution $x^*(\cdot):[t_0, t_1^*] \to R^n$ of (13.141)–(13.142), such that

$$\int_{t_0}^{t_1^*} \hat{f}_0[x^*(t), w^*, u^*(t)] \, dt \leq \int_{t_0}^{t_1} \hat{f}_0[x(t), w, u(t)] \, dt$$

for all admissible $u(\cdot):[t_0, t_1] \to R^m$ and all $w \in R^\gamma$, with the corresponding solution $x(\cdot):[t_0, t_1] \to R^n$ of (13.141) satisfying (13.142); of course, if the initial state is not prescribed, there may be more than one such solution.

This problem is readily cast into the form of the problem posed in Chapter 10, treated in Chapter 11, and generalized in Section 13.2. To accomplish this we consider the augmented state

$$\tilde{x} \triangleq [x_1 \ x_2 \cdots x_{n+\gamma}]^T$$

and let $\tilde{f}_i(\cdot): R^{n+\gamma} \times R^m \to R^1$ be such that for $x_{n+i} \triangleq w_i$

$$\tilde{f}_i(\tilde{x}, u) = \hat{f}_i(x, w, u), \quad i = 0, 1, \ldots, n,$$
$$\tilde{f}_i(\tilde{x}, u) = 0, \quad i = n+1, n+2, \ldots, n+\gamma.$$

The state equations for the augmented system are

$$\dot{x}_i(t) = \tilde{f}_i[\tilde{x}(t), u(t)], \quad i = 1, 2, \ldots, n+\gamma. \tag{13.144}$$

The end conditions are

$$\tilde{x}(t_0) \in \tilde{\theta}^0, \quad \tilde{x}(t_1) \in \tilde{\theta}^1$$

where

$$\tilde{\theta}^0 \triangleq \{\tilde{x} \in R^{n+\gamma} | x \in \theta^0\},$$
$$\tilde{\theta}^1 \triangleq \{\tilde{x} \in R^{n+\gamma} | x \in \theta^1\}; \tag{13.145}$$

that is, the end values of the $x_i(\cdot)$, $i = n+1, n+2, \ldots, n+\gamma$, are *not* prescribed.

The cost for the augmented system is

$$\int_{t_0}^{t_1} \tilde{f}_0[\tilde{x}(t), u(t)] \, dt. \tag{13.146}$$

Since the first n components of the state are governed by the same equations and are subject to the same end conditions in both versions, while the last γ state components are simply constants in the latter version, in place of constant parameters in the original one, the systems (13.141)–(13.143) and (13.144)–(13.146) are equivalent to one another. In other words, a choice of $u(\cdot)$ and w in the former, and of $u(\cdot)$ and $[x_{n+1}(t)\; x_{n+2}(t) \cdots x_{n+\gamma}(t)]^T \equiv w$ in the latter, yield the same $x(\cdot)$ for given $x(t_0) \in \theta^0$ in both versions, and hence the same values of the respective costs. However, the problem of minimizing the cost for the system (13.144)–(13.146) is one of optimal control on $\tilde{\theta}^0$, that is, of determining a control *and* an initial state resulting in a transfer to $\tilde{\theta}^1$ and the minimum value of the cost. Thus the necessary conditions which apply to the system (13.144)–(13.146)—namely, Theorem 11.1 and the initial transversality condition—are also necessary for the optimality of a control *and* of a parameter in the system (13.141)–(13.143).

Let

$$\tilde{y} \triangleq [x_0 \; x_1 \cdots x_{n+\gamma}]^T, \qquad \tilde{\lambda} \triangleq [\lambda_0 \; \lambda_1 \cdots \lambda_{n+\gamma}]^T,$$

and define

$$\tilde{H}(\tilde{\lambda}, \tilde{y}, u) \triangleq \sum_{i=0}^{n+\gamma} \lambda_i \tilde{f}_i(\tilde{x}, u).$$

Then the adjoint equations for the augmented system are

$$\dot{\lambda}_i(t) = -\frac{\partial \tilde{H}[\tilde{\lambda}(t), \tilde{y}^*(t), u^*(t)]}{\partial x_i}, \qquad i = 0, 1, \ldots, n+\gamma. \quad (13.147)$$

The conditions of Theorem 11.1 become

(a) $\max_{u \in U} \tilde{H}[\tilde{\lambda}(t), \tilde{y}(t), u] = \tilde{H}[\tilde{\lambda}(t), \tilde{y}^*(t), u^*(t)]$,

(b) $\tilde{H}[\tilde{\lambda}(t), \tilde{y}^*(t), u^*(t)] = 0$,

(c) $\lambda_0(t) = \text{constant} \leq 0$.

The initial transversality condition (13.8)–(13.9) implies

$$\sum_{i=1}^{n} \lambda_i(t_0) \eta_i^0 = 0 \qquad (13.148)$$

subject to

$$\sum_{i=1}^{n} \frac{\partial \theta_j^0(x^*(t_0))}{\partial x_i} \eta_i^0 = 0, \qquad j=1,2,\ldots,p, \qquad (13.149)$$

as well as

$$\lambda_{n+i}(t_0) = 0, \qquad i=1,2,\ldots,\gamma. \qquad (13.150)$$

The terminal transversality condition (11.38)–(11.39) results in

$$\sum_{i=1}^{n} \lambda_i(t_1^*) \eta_i^1 = 0 \qquad (13.151)$$

subject to

$$\sum_{i=1}^{n} \frac{\partial \theta_j^1(x^*(t_1^*))}{\partial x_i} \eta_i^1 = 0, \qquad j=1,2,\ldots,q, \qquad (13.152)$$

as well as

$$\lambda_{n+i}(t_1^*) = 0, \qquad i=1,2,\ldots,\gamma. \qquad (13.153)$$

Now let $\hat{H}(\cdot): R^{n+1} \times R^{n+1} \times R^\gamma \times R^m \to R^1$ be defined by

$$\hat{H}(\lambda, y, w, u) \triangleq \sum_{i=0}^{n} \lambda_i \hat{f}_i(x, w, u).$$

Then, in view of the state equations (13.144), one has

$$\tilde{H}(\tilde{\lambda}, \tilde{y}, u) = \hat{H}(\lambda, y, w, u)$$

so that the first $n+1$ of the adjoint equations (13.147) become

$$\dot{\lambda}_i(t) = -\frac{\partial \hat{H}[\lambda(t), y^*(t), w^*, u^*(t)]}{\partial x_i}, \qquad i=0,1,\ldots,n. \quad (13.154)$$

Furthermore, as a consequence of (13.150) and (13.153), the integration of the remaining adjoint equations yields

$$\int_{t_0}^{t_1^*} \frac{\partial \hat{H}[\lambda(t), y^*(t), w^*, u^*(t)]}{\partial w_i} dt = 0, \qquad i=1,2,\ldots,\gamma. \quad (13.155)$$

If $\lambda(t_0)=0$ or $\lambda(t_1^*)=0$, then it follows from (13.150) and (13.153) that $\tilde{\lambda}(t_0)=0$ or $\tilde{\lambda}(t_1^*)=0$. But we require a nonzero solution, $\tilde{\lambda}(\cdot)$, of (13.147). Thus neither $\lambda(t_0)$ nor $\lambda(t_1^*)=0$; consequently, $\lambda(t)\neq 0$ for all $t\in[t_0,t_1^*]$.

We can now summarize the necessary conditions for optimality in the original system (13.141)–(13.143).

Theorem 13.2. *If the control* $u^*(\cdot):[t_0,t_1^*]\to R^m$ *is optimal on* θ^0 *and* w^* *is an optimal parameter, with the corresponding solution* $x^*(\cdot):[t_0,t_1^*]\to R^n$, *then there exists a nonzero solution* $\lambda(\cdot):[t_0,t_1^*]\to R^{n+1}$ *of the adjoint equations* (13.154) *such that*

(a) $\max\limits_{u\in U} \hat{H}[\lambda(t),y^*(t),w^*,u] = \hat{H}[\lambda(t),y^*(t),w^*,u^*(t)]$,

(b) $\hat{H}[\lambda(t),y^*(t),w^*,u^*(t)]=0$,

(c) $\lambda_0(t)=$ constant ≤ 0,

for all $t\in[t_0,t_1^*]$, *with* (a) *and* (b) *holding for* $u^*(\bar{t}-0)$ *and* $u^*(\bar{t}+0)$ *if* $u^*(\cdot)$ *is discontinuous at* $\bar{t}\in(t_0,t_1^*)$,

(d) $\displaystyle\int_{t_0}^{t_1^*} \frac{\partial \hat{H}[\lambda(t),y^*(t),w^*,u^*(t)]}{\partial w_i}\,dt = 0, \quad i=1,2,\ldots,\gamma$,

and

(e) *the transversality conditions* (13.148)–(13.149) *and* (13.151)–(13.152) *are satisfied.*

Note that the conditions (a)–(c) are the same as those of Theorem 11.1 except for the dependence of $\hat{H}(\cdot)$ on the parameter w. The transversality conditions are also unchanged. However now there are γ additional conditions which must be met by the optimal parameters w_i^*, $i=1,2,\ldots,\gamma$.

If a control $u(\cdot):[t_0,t_1]\to R^m$ and a parameter $w\in R^\gamma$ are such that the conditions of Theorem 13.2 are met, we call the control-parameter pair *extremal*.

Finally it should be noted that the various generalizations discussed in Sections 13.4, 13.6, 13.7, 13.10, and 13.12 are also applicable to the problem of the optimal control and parameter choice. The consequent modifications in the conditions of Theorem 13.2 are readily deduced by considering the relevant results for the equivalent system (13.144)–(13.146); this is left as an exercise for the interested reader. For instance, if the interval $[t_0,t_1]$ is prescribed, then the condition (b) of Theorem 13.2 is replaced by

$$\hat{H}[\lambda(t),y^*(t),w^*,u^*(t)]=\text{constant}.$$

13.15. An Illustrative Example

To illustrate the results of Section 13.14 let us consider a simple system with the state equation

$$\dot{x}_1(t) = u(t) + w. \tag{13.156}$$

The end states and the interval are given; that is,

$$x_1(t_0) = x_1^0, \quad x_1(t_1) = x_1^1, \quad t_1 - t_0 = \tau_1 > 0. \tag{13.157}$$

The control is constrained by

$$|u(t)| \leq 1, \tag{13.158}$$

and the cost is

$$\int_{t_0}^{t_1} u^2(t) \, dt. \tag{13.159}$$

Before utilizing the necessary conditions of Theorem 13.2, as modified by the prescription of the interval, in order to deduce an extremal control-parameter pair, we note that the solution of the problem is an obvious one; that is, the optimal control $u^*(\cdot):[t_0, t_1] \to R^1$ and parameter $w^* \in R^1$ can be deduced by inspection. The minimum value of the cost (13.159) is zero so that

$$u^*(t) \equiv 0. \tag{13.160}$$

The corresponding parameter value is obtained by integrating (13.156) subject to (13.157); namely,

$$w^* = \frac{x_1^1 - x_1^0}{\tau_1}. \tag{13.161}$$

Now let us employ Theorem 13.2 to derive an extremal control and parameter, $u(\cdot)$ and w. By (13.156) and (13.159) we have

$$\hat{H}(\lambda, y, w, u) = \lambda_0 u^2 + \lambda_1 (u + w)$$

so that

$$\dot{\lambda}_1(t) = 0$$

and hence

$$\lambda_1(t) \equiv \text{constant}. \tag{13.162}$$

However the condition (d) of Theorem 13.2 leads to

$$\int_{t_0}^{t_1} \lambda_1(t) \, dt = 0$$

whence, in view of (13.162), we conclude that

$$\lambda_1(t) \equiv 0. \tag{13.163}$$

Thus, to assure a nonzero solution of the adjoint equations, $\lambda_0(t) \equiv \text{constant} \neq 0$; that is, as a result of the condition (c) of Theorem 13.2, it follows that

$$\lambda_0(t) \equiv \text{constant} < 0. \tag{13.164}$$

Then the condition (a) of the theorem, together with (13.163) and (13.164), implies that

$$u(t) = 0 \quad \forall t \in [t_0, t_1]$$

so that again the only parameter value for which the end conditions are met is

$$w = \frac{x_1^1 - x_1^0}{T_1}.$$

13.16. End Point Inequality Constraints

Here we examine briefly the effect of imposing an inequality constraint on the terminal or initial value of a state component or of the independent variable.

In particular suppose that the target set is a smooth manifold defined by

$$\theta^1 \triangleq \{ x \in R^n \mid \theta_i^1(x) = 0, \, i = 1, 2, \ldots, q < n \}$$

where the $\theta_i^1(\cdot)$ are constant with respect to x_k; that is,

$$\frac{\partial \theta_i^1(x)}{\partial x_k} \equiv 0, \quad i = 1, 2, \ldots, q. \tag{13.165}$$

However the terminal value of x_k is constrained by

$$x_{k \min}^1 \leq x_k \leq x_{k \max}^1, \tag{13.166}$$

where $x^1_{k\min}$ and $x^1_{k\max} > x^1_{k\min}$ are given constants.

We must allow for two possibilities:

(i) The solution $x^*(\cdot):[t_0, t_1^*] \to R^n$ generated by the optimal control $u^*(\cdot):[t_0, t_1^*] \to R^m$ is such that

$$x^1_{k\min} < x_k^*(t_1^*) < x^1_{k\max},$$

that is, the terminal value of $x_k(\cdot)$ is (locally) unconstrained. In this event the terminal transversality condition implies that

$$\lambda_k(t_1^*) = 0.$$

(ii) The solution generated by the optimal control is such that

$$x_k^*(t_1^*) = x^1_{k\min} \quad \text{or} \quad x_k^*(t_1^*) = x^1_{k\max}.$$

When deducing *extremal* controls from the necessary conditions one must allow for both of these possibilities.

If the target set is

$$\theta^1 \triangleq \{x \in R^n | \theta_i^1(x) = 0, \ i = 1, 2, \ldots, q \leq n\}$$

and the terminal value of the independent variable, t_1, is constrained by

$$t_{1\min} \leq t \leq t_{1\max} \tag{13.167}$$

where $t_{1\min}$ and $t_{1\max} > t_{1\min}$ are given constants, one is again faced with two possibilities:

(i) The optimal control $u^*(\cdot):[t_0, t_1^*] \to R^m$ is such that

$$t_{1\min} < t_1^* < t_{1\max}$$

so that t_1 is (locally) unconstrained. Then, provided the state equations are autonomous and the cost integrand does not depend explicitly on t, the condition (b) of Theorem 11.1 applies; namely,

$$H[\lambda(t), y^*(t), u^*(t)] = 0 \qquad \forall t \in [t_0, t_1^*].$$

If the state equations are not autonomous or the cost integrand depends explicitly on t, then the terminal transversality condition (13.51)–(13.52) of

Section 13.6 applies; it implies that

$$\hat{H}[\lambda(t_1^*), y^*(t_1^*), t_1^*, u^*(t_1^*)] = 0.$$

(ii) The optimal control is such that

$$t_1^* = t_{1\min} \quad \text{or} \quad t_1^* = t_{1\max}.$$

In this event the relevant terminal transversality condition (13.51)–(13.52) applies with

$$\hat{\theta}_{q+1}^1(x, t) = t - t_{1\min} (\text{or } t_{1\max}) = 0.$$

Entirely analogous results hold for initial value inequality constraints.

Finally we note that the discussion of this section is directly applicable to inequality constraints imposed on a parameter w_k for systems of the type treated in Section 13.14. Suppose that w_k is constrained by

$$w_{k\min} \leq w_k \leq w_{k\max} \tag{13.168}$$

where $w_{k\min}$ and $w_{k\max} > w_{k\min}$ are given constants. In view of the arguments used in Section 13.14 we have $x_{n+k}(t) \equiv w_k$, so that the constraint (13.168) and the constraint (13.166) for x_{n+k} are equivalent to one another. Thus again we have two possibilities:

(i) The optimal control and parameter are such that

$$w_{k\min} < w_k^* < w_{k\max},$$

so that w_k is (locally) unconstrained; in this event, condition (d) of Theorem 13.2 applies for $i = k$.

(ii) The optimal control and parameter are such that

$$w_k^* = w_{k\min} \quad \text{or} \quad w_k^* = w_{k\max},$$

replacing condition (d) for $i = k$.

13.17. An Illustrative Example

At the conclusion of the last section we mentioned the possibilities which must be taken into account when deducing an extremal control and parameter in the presence of parameter bounds. To illustrate this consider a system with the state

Chap. 13 • Some Generalizations

equations

$$\dot{x}_1(t) = x_2(t) + w, \qquad \dot{x}_2(t) = u(t). \tag{13.169}$$

The end states are prescribed; namely,

$$x_1(t_0) = x_1^0, \qquad x_2(t_0) = x_2^0,$$
$$x_1(t_1) = 0, \qquad x_2(t_1) = 0.$$

The control and the parameter, respectively, are constrained by

$$|u(t)| \le 1$$

and

$$|w| \le W > 0$$

where W is a given constant.

The cost is

$$t_1 - t_0 = \int_{t_0}^{t_1} dt, \tag{13.170}$$

that is, if t stands for time, we seek a control function and a parameter value which result in the transfer from x^0 to $x^1 = 0$ in minimum time. Here we are content with employing Theorem 13.2 to deduce an *extremal* control and parameter pair, $u(\cdot): [t_0, t_1] \to R^1$ and $w \in R^1$.

In view of (13.169) and (13.170) we have

$$\hat{H}(\lambda, y, w, u) = \lambda_0 + \lambda_1(x_2 + w) + \lambda_2 u \tag{13.171}$$

so that the adjoint equations (13.154) become

$$\dot{\lambda}_0(t) = 0,$$
$$\dot{\lambda}_1(t) = 0, \qquad \dot{\lambda}_2(t) = \lambda_1(t)$$

whence, in particular, for $t \in [t_0, t_1]$

$$\lambda_1(t) - c_1, \qquad \lambda_2(t) = c_2 - c_1 t \triangleq \sigma(t) \tag{13.172}$$

where c_1 and c_2 are constants. Also, the condition (c) of Theorem 13.2 applies; that is,

$$\lambda_0(t) \equiv \text{constant} \le 0.$$

We note at once that not both c_1 and c_2 can be zero lest $\lambda_0(t) \equiv 0$ as well; this follows from the condition (b) of Theorem 13.2. Hence the *switching function* $\sigma(\cdot)$, defined in (13.172), can have at most one zero on $[t_0, t_1]$, and so the condition (a) of the theorem implies that the extremal control $u(\cdot)$ is bang–bang; namely,

$$u(t) = 1 \quad \text{for } \sigma(t) > 0,$$
$$u(t) = -1 \quad \text{for } \sigma(t) < 0,$$

and there can be at most one switch.

Concerning the extremal parameter w we must allow for three possibilities:
(a) $w = W$,
(b) $w = -W$,
(c) $|w| < W$.

Let us consider these three cases in turn.

Case (a). With $w = W$ and $u(t) \equiv 1$ or -1, the integration of the state equations (13.169) yields

$$x_1(t) = \tfrac{1}{2}[x_2(t) + W]^2 + k_+ \quad \text{for } u(t) \equiv 1,$$
$$x_1(t) = -\tfrac{1}{2}[x_2(t) + W]^2 + k_- \quad \text{for } u(t) \equiv -1$$

where k_+ and k_- are constants. For given values of these constants the trajectories are parabolas; these are illustrated in Figures 13.6 and 13.7, respectively, with arrows indicating the direction of increasing t.

Since there can be at most one switch in the value of an extremal control, a corresponding trajectory consists of at most two parabolic arcs, one of which ends at the terminal state; these terminal arcs belong to the parabolas whose equations are

$$x_1 = \tfrac{1}{2}x_2^2 + Wx_2, \qquad x_1 = -\tfrac{1}{2}x_2^2 - Wx_2, \qquad (13.173)$$

respectively. They are denoted by AO and BO in Figures 13.6 and 13.7. The curve AOB is the *switching curve*; that is, if $x(t_s) \in AOB$, $t_s \in (t_0, t_1)$, then $u(\cdot)$ is discontinuous at $t = t_s$. If the initial state, x^0, lies in the region *above* AOB, then the

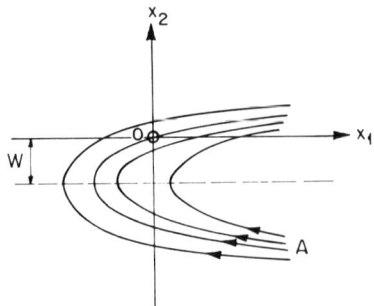

Figure 13.6. Trajectories with $w = W$ and $u(t) \equiv 1$.

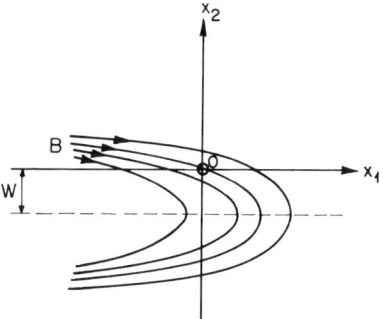

Figure 13.7. Trajectories with $w = W$ and $u(t) \equiv -1$.

switching sequence is $-1 \to 1$; that is,

$$u(t) \equiv -1 \quad \text{for } t \in [t_0, t_s),$$
$$u(t) \equiv 1 \quad \text{for } t \in [t_s, t_1]. \tag{13.174}$$

Conversely, if x^0 is in the region *below AOB*, then the switching sequence is $1 \to -1$; that is,

$$u(t) \equiv 1 \quad \text{for } t \in [t_0, t_s),$$
$$u(t) \equiv -1 \quad \text{for } t \in [t_s, t_1]. \tag{13.175}$$

This is illustrated in Figure 13.8.

The value of the cost, $t_1 - t_0$, is obtained from the integration of the state equations; namely,

$$T_a \triangleq t_1 - t_0 = |x_2^0 - 2x_2(t_s)_a|, \tag{13.176}$$

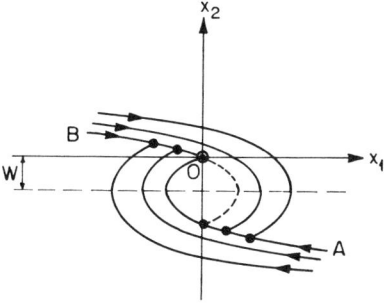

Figure 13.8. Switching sequence for $w = W$.

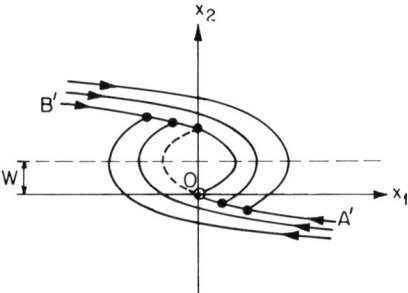

Figure 13.9. Switching sequence for $w = -W$.

where

$$2x_2(t_s)_a = -2W - \left[4W^2 + 2(x_2^0)^2 + 4Wx_2^0 + 4x_1^0\right]^{1/2}$$

for the switching sequence (13.174), and

$$2x_2(t_s)_a = -2W + \left[4W^2 + 2(x_2^0)^2 + 2Wx_2^0 - 4x_1^0\right]^{1/2}$$

for the switching sequence (13.175).

Case (b). With $w = -W$ and bang–bang control we proceed as in case (a) and obtain analogous results. In particular, the switching curve denoted by $A'OB'$ in Figure 13.9 consists of two half parabolas whose equations are

$$x_1 = \tfrac{1}{2}x_2^2 - Wx_2, \qquad x_1 = -\tfrac{1}{2}x_2^2 + Wx_2, \qquad (13.177)$$

respectively.

The value of the cost is

$$T_b \triangleq t_1 - t_0 = |-x_2^0 + 2x_2(t_s)_b|, \qquad (13.178)$$

where

$$2x_2(t_s)_b = 2W - \left[4W^2 + 2(x_2^0)^2 - 4Wx_2^0 + 4x_1^0\right]^{1/2}$$

for x^0 lying *above* $A'OB'$, corresponding to the switching sequence $-1 \to 1$, and

$$2x_2(t_s)_b = 2W + \left[4W^2 + 2(x_2^0)^2 - 4Wx_2^0 - 4x_1^0\right]^{1/2}$$

for x^0 belonging to the region *below* $A'OB'$, corresponding to the switching sequence $1 \to -1$.

Chap. 13 • Some Generalizations

Case (c). If $|w| < W$, then the condition (d) of Theorem 13.2 must be satisfied. This condition, together with (13.172), implies that $c_1 = 0$ so that $\sigma(t) \equiv c_2$. Consequently there is no switch in the extremal control and

$$u(t) \equiv 1 \quad \text{if } c_2 > 0,$$
$$u(t) \equiv -1 \quad \text{if } c_2 < 0.$$

Then, on integrating the state equations, one obtains

$$\left.\begin{array}{l} u(t) \equiv -1 \\ w = -\dfrac{x_1^0}{x_2^0} - \dfrac{1}{2} x_2^0 \triangleq w_c \\ t_1 - t_0 = x_2^0 \triangleq \tau_c \end{array}\right\} \quad \text{for } x_2^0 > 0,$$

$$\left.\begin{array}{l} u(t) \equiv 1 \\ w = \dfrac{x_1^0}{x_2^0} - \dfrac{1}{2} x_2^0 \triangleq w_c \\ t_1 - t_0 = -x_2^0 \triangleq \tau_c \end{array}\right\} \quad \text{for } x_2^0 < 0.$$

(13.179)

However, w_c in (13.179) is constrained by $|w_c| < W$. Consequently case (c) is possible only for initial states such that

$$x_1^0 = \tfrac{1}{2}(x_2^0)^2 + w x_2^0, \qquad x_2^0 < 0,$$
$$x_1^0 = -\tfrac{1}{2}(x_2^0)^2 - w x_2^0, \qquad x_2^0 > 0,$$

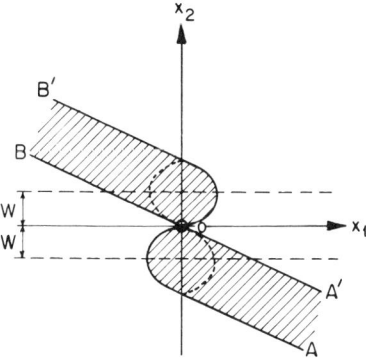

Figure 13.10. Initial state region for $|w| < W$.

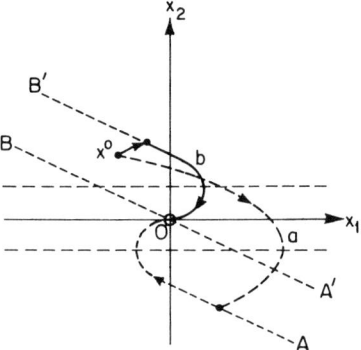

Figure 13.11. Trajectories for cases (a) and (b).

where $|w| < W$. Then, in view of (13.173) and (13.177), the initial state must lie in the open region bounded by AOB and $A'OB'$; this is shown in Figure 13.10.

Now let us compare the values of the cost, $t_1 - t_0$, corresponding to the three cases. Suppose first that the initial state lies in the region for which case (c) is possible. For $x_2^0 > 0$ it follows from (13.176), (13.178), and (13.179), together with an analysis of Figure 13.11, that

$$\tau_a = x_2^0 - 2x_2(t_s)_a, \quad x_2(t_s)_a < 0,$$
$$\tau_b = 2x_2(t_s)_b - x_2^0, \quad x_2(t_s)_b > x_2^0,$$
$$\tau_c = x_2^0.$$

Thus, for a given initial state, we conclude that

$$\tau_c < \tau_a \text{ and } \tau_b$$

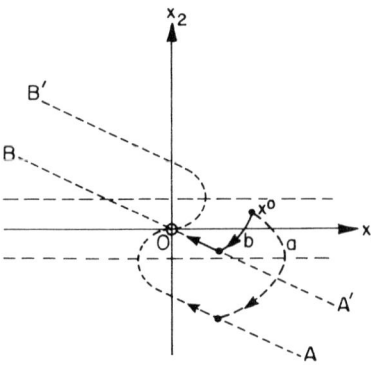

Figure 13.12. Trajectories for cases (a) and (b).

Chap. 13 • Some Generalizations

Table 13.1. Candidates for Optimal Control and Parameter

Initial state	Switching sequence	Parameter
Below AOB	$1 \to -1$	W
Above $A'OB'$	$-1 \to 1$	$-W$
Between AO and $A'O$	1	$\dfrac{x_1^0}{x_2^0} - \dfrac{1}{2} x_2^0$
Between BO and $B'O$	-1	$-\dfrac{x_1^0}{x_2^0} - \dfrac{1}{2} x_2^0$

so that, provided case (c) is possible, it is "better" than cases (a) and (b); that is, the extremal control and parameter corresponding to case (a) or case (b) cannot be optimal. The same conclusion is reached for $x_2^0 < 0$.

Next we compare the costs τ_a and τ_b for the initial state lying *outside* the region for which case (c) is possible. If x^0 lies *above* $A'OB'$, as shown in Figure 13.12, then

$$\tau_a = x_2^0 - 2x_2(t_s)_a,$$
$$\tau_b = x_2^0 - 2x_2(t_s)_b,$$

and

$$x_2(t_s)_a < x_2(t_s)_b.$$

Consequently we have

$$\tau_b < \tau_a.$$

Similarly, if x^0 lies *below* AOB, one can conclude that

$$\tau_a < \tau_b.$$

In the light of the results obtained, the *extremal* control-parameter pairs which are candidates for *optimal* ones are listed in Table 13.1.

Exercises

13.1. Find the plane curve of minimum length which joins a point on the curve

$$\theta^0 = \{x \in R^2 \mid x_1^2 - x_2 = 0\}$$

to a point on

$$\theta^1 = \{x \in R^2 \mid x_1 + x_2 + 3 = 0\}.$$

13.2. Recall the discussion of Section 13.4. Show that $u^*(\cdot)$ is optimal on θ^0 for the original system with the cost (13.11) if $\tilde{u}^*(\cdot)$ is optimal on $\tilde{\theta}^0$ for the augmented system with the cost (13.16).

13.3. Consider the state equations
$$\dot{x}_1(t) = x_2(t), \qquad \dot{x}_2(t) = u(t)$$
with the control constraint
$$|u(t)| \leq 1.$$
It is desired to transfer the state from the given initial state x^0 to one in the target set
$$\theta^1 = \{x \in R^2 \mid x_2 = 0\},$$
while minimizing the value of the cost
$$\int_{t_0}^{t_1} dt + \alpha x_1^2(t_1)$$
where $\alpha > 0$ is a given constant. Discuss extremal control.

13.4. Consider the system treated in Section 13.5.
(a) Discuss extremal control corresponding to $\lambda_i(t) \equiv 0$, $i = 1, 2, \ldots, 4$.
(b) Discuss the possibility of $\lambda_i(t) \equiv 0$, $i = 1, 2, \ldots, 5$. Can this happen? If so, what does it mean?

13.5. Recall the system discussed in Section 13.5. Suppose that the initial state is prescribed and that the target set is
$$\theta^1 = \{x \in R^5 \mid x_5 = x_5^1 < x_5^0\}.$$
It is desired to maximize the terminal value of $x_3(t)$. Discuss extremal control.

13.6. Recall the problem of Section 13.5. Show that the switching sequence
$$u_3 = 0 \rightarrow u_3 = u_3^{\max} \rightarrow u_3 = 0$$
cannot be extremal.

13.7. The following state equations describe the motion of a power-limited rocket in planar flight:
$$\dot{x}_1(t) = x_3(t),$$
$$\dot{x}_2(t) = x_4(t),$$
$$\dot{x}_3(t) = \frac{u_1(t) u_3(t) u_4(t)}{x_5(t)},$$
$$\dot{x}_4(t) = \frac{u_2(t) u_3(t) u_4(t)}{x_5(t)},$$
$$\dot{x}_5(t) = -u_4(t).$$

Chap. 13 • Some Generalizations

The control constraint set, U, is defined by

$$u_1^2 + u_2^2 = 1,$$
$$0 \leq u_3^2 u_4 \leq \alpha,$$
$$0 \leq u_3 \leq \beta,$$
$$0 \leq u_4 \leq \gamma,$$

where α, β, and γ are given positive constants. Discuss extremal control for the case $\beta^2 \gamma > \alpha$. In particular show that an extremal control must have its value on the boundary of U if the cost is a terminal one.

13.8. Consider the system whose state equation is

$$\dot{x}_1(t) = u(t)$$

with the control constrained by

$$|u(t)| \leq 1.$$

The state is to be transferred from $x_1(t_0) = x_1^0$ to $x_1(t_1) = x_1^1$ in the given interval $t_1 - t_0 = \tau_1$. The cost to be minimized is

$$\int_{t_0}^{t_1} x_1^2(t)\, dt.$$

Discuss extremal control.

13.9. Recall the problem treated in Section 13.11. Show that an appropriate change of control variable allows one to reduce the problem to an equivalent one with a *constant* control constraint set. Discuss extremal control.

13.10. Recall the problem treated in Section 13.11. Compute the values of the switching time, t_s, and of the transfer time, $t_1 - t_0$, corresponding to the extremal control.

13.11. The state equations of a rocket plane in horizontal flight over the surface of an airless planet are

$$\dot{x}_1(t) = \left\{ \left[\frac{cu_3(t)}{x_3(t)}\right]^2 - g^2 \right\}^{1/2} u_1(t),$$

$$\dot{x}_2(t) = \left\{ \left[\frac{cu_3(t)}{x_3(t)}\right]^2 - g^2 \right\}^{1/2} u_2(t),$$

$$\dot{x}_3(t) = -u_3(t),$$

where c and g are given positive constants. The control constraint set, $\hat{U}(x)$, is defined by

$$u_1^2 + u_2^2 = 1,$$

$$\frac{g}{c} x_3 \leq u_3 \leq u_3^{max}$$

where u_3^{max} is a given positive constant. It is desired to transfer the state from a given initial one, x^0, to

$$\theta^1 = \{x \in R^3 | x_1 = x_1^1, x_2 = x_2^1\},$$

while maximizing the terminal value of $x_3(t)$. Discuss extremal control.

13.12. The state equations for a rocket in rectilinear motion are

$$\dot{x}_1(t) = x_2(t),$$

$$\dot{x}_2(t) = \frac{wu(t)}{x_3(t)},$$

$$\dot{x}_3(t) = -u(t).$$

The constraints on the control value $u(t)$ and on the constant parameter w are

$$0 \leq u(t) \leq u^{max},$$

$$0 \leq w \leq w^{max},$$

where u^{max} and w^{max} are given positive constants. It is desired to transfer the state from the given initial one, x^0, to

$$\theta^1 = \{x \subset R^3 | x_1 = x_1^1, x_2 = x_2^1\},$$

while maximizing the terminal value of $x_3(t)$. Discuss extremal control and parameter choice.

13.13. Consider the problem discussed in Section 13.8. However, now suppose that the transfer time, $t_1 - t_0$, is not prescribed, but rather that an upper bound is placed on it; that is,

$$t_1 - t_0 \leq \tau_1$$

where $\tau_1 > 0$ is given. Discuss extremal control.

13.14. Consider the system described by the state equations

$$\dot{x}_1(t) = x_2(t), \qquad \dot{x}_2(t) = u(t)$$

Chap. 13 • Some Generalizations

and subject to the isoperimetric constraint

$$\int_{t_0}^{t_1} u^2(t)\, dt \le l$$

where $l > 0$ is given. It is desired to transfer the state from the given initial one, x^0, to the given terminal one, $x^1 = 0$, while minimizing the value of the interval

$$t_1 - t_0 = \int_{t_0}^{t_1} dt.$$

Discuss extremal control.

13.15. Consider again the system of Exercise 13.14. However, now only the initial state, x^0, is given and the interval $t_1 - t_0 = \tau_1$ is prescribed. It is desired to minimize the value of

$$\int_{t_0}^{t_1} \left[x_1^2(t) + x_2^2(t) \right] dt.$$

Discuss extremal control.

13.16. Recall the discussion of end point inequality constraints in Section 13.16. Using a method analogous to that employed in the proof of Theorem 13.1, Section 13.12, show that

$$x_k^*(t_1^*) = x_{k\,\min}^1 \Rightarrow \lambda_k(t_1^*) \ge 0,$$
$$x_k^*(t_1^*) = x_{k\,\max}^1 \Rightarrow \lambda_k(t_1^*) \le 0.$$

13.17. Recall the problem treated in Section 13.17. Use the result of Exercise 13.16 to rederive the extremal control for initial states lying below AOB and above $A'OB'$, respectively.

13.18. Using a method analogous to that employed in the proof of Theorem 13.1, Section 13.12, derive the results of Section 13.10. Show also that $\nu_i(t) \le 0$, $i = l+1, l+2, \ldots, l+r$.

14

Special Systems

14.1. Introduction

In this chapter we shall discuss various properties of an extremal control for systems of a special kind. We shall begin the discussion with so-called linear time-invariant state equations and componentwise control constraints. For such systems we introduce the notion of a *switching function*, already encountered earlier in connection with particular problems, whose behavior determines the nature of an extremal control. We continue with a treatment of "time-optimal" control and deduce some of its special properties; these are then illustrated in some simple examples.

The next special system to be considered is one in which some of the control components enter linearly into the state equations. For such systems the behavior of the switching functions is again of interest. In particular, there arises the concept of the so-called *singular* control. We conclude the chapter with some applications.

14.2. Linear Time-Invariant State Equations

Consider the state equations

$$\dot{x}_i(t) = \sum_{j=1}^{n} a_{ij} x_j(t) + \sum_{k=1}^{m} b_{ik} u_k(t), \qquad i = 1, 2, \ldots, n, \qquad (14.1)$$

where the coefficients a_{ij} and b_{ik} are given constants. Alternatively we may write (14.1) in vector-matrix form as

$$\dot{x}(t) = Ax(t) + Bu(t) \tag{14.2}$$

where A and B are constant matrices of dimension $n \times n$ and $n \times m$, respectively.

We suppose further that the control constraint set is a parallelepiped in R^m; that is,

$$U = \{u \in R^m | u_k^{\min} \leq u_k \leq u_k^{\max}, k = 1, 2, \ldots, m\} \tag{14.3}$$

where the $u_k^{\min}$ and $u_k^{\max}$ are given constants.

Finally we restrict our discussion to linear cost integrands; namely,

$$f_0(x, u) = \sum_{j=1}^{n} a_{0j} x_j + \sum_{k=1}^{m} b_{0k} u_k + c_0$$

so that the cost is

$$\int_{t_0}^{t_1} \left[\sum_{j=1}^{n} a_{0j} x_j(t) + \sum_{k=1}^{m} b_{0k} u_k(t) + c_0 \right] dt \tag{14.4}$$

where the a_{0j}, b_{0k}, and c_0 are given constants.

As before we seek an admissible control $u^*(\cdot):[t_0, t_1^*] \to R^m$ with a corresponding solution $x^*(\cdot):[t_0, t_1^*] \to R^n$ such that the state is transferred between prescribed smooth manifolds θ^0 and θ^1 while the value of the cost is minimized; that is, we desire a control which is optimal on θ^0. In the ensuing discussion we shall employ the necessary conditions derived earlier, especially Theorem 11.1, to deduce some properties of an *extremal* control.

14.3. The Switching Function

When using the maximum principle, Theorem 11.1, one has need of the function $H(\cdot)$. For the type of problem posed in Section 14.2 it is given by

$$H(\lambda, y, u) = \lambda_0 c_0 + \sum_{i=0}^{n} \lambda_i \left[\sum_{j=1}^{n} a_{ij} x_j + \sum_{k=1}^{m} b_{ik} u_k \right]. \tag{14.5}$$

Now suppose that the control $u(\cdot):[t_0, t_1] \to R^m$, with the corresponding solution $x(\cdot): [t_0, t_1] \to R^n$ of the state equations (14.1) and $\lambda(\cdot):[t_0, t_1]\to R^{n+1}$ of the adjoint equations (11.28), is such that $x(t_0) \in \theta^0$, $x(t_1) \in \theta^1$ and the conditions of Theorem 11.1 are satisfied; that is, $u(\cdot)$ is an *extremal* control and hence a candidate for an optimal one on θ^0.

Since the coefficient of u_k in (14.5) is $\sum_{i=0}^n \lambda_i b_{ik}$, we introduce the function $\sigma_k(\cdot):[t_0, t_1]\to R^1$ defined by

$$\sigma_k(t) \triangleq \sum_{i=0}^n \lambda_i(t) b_{ik}. \tag{14.6}$$

Then it follows from the condition (a) of Theorem 1.1, together with (14.3), that

$$\begin{aligned} u_k(t) &= u_k^{\min} && \text{if } \sigma_k(t) < 0, \\ u_k(t) &= u_k^{\max} && \text{if } \sigma_k(t) > 0. \end{aligned} \tag{14.7}$$

In other words, if $\sigma_k(t) \neq 0$ then $u_k(t)$ takes on one of its limiting values, $u_k^{\min}$ or $u_k^{\max}$, and if $\sigma_k(\cdot)$ changes its sign then $u_k(\cdot)$ *switches* from one of its limiting values to the other in accordance with (14.7). For this reason, $\sigma_k(\cdot)$ is called the *switching function* for the kth control component. Since the adjoint variable $\lambda(\cdot)$ is continuous on $[t_0, t_1]$, so is the switching function $\sigma_k(\cdot)$. Consequently, $\sigma_k(\cdot)$ must pass through zero if it changes sign. Thus, if $\sigma_k(\cdot)$ changes sign at $t_s \in (t_0, t_1)$, then

$$\sigma_k(t_s) = 0$$

and $u_k(\cdot)$ is discontinuous at $t = t_s$. Clearly the behavior of $\sigma_k(\cdot)$ is of great interest. In the following sections we deduce some properties of the switching function for the case of "time-optimal" control.

14.4. Time-Optimality and Bang–Bang Control

If the switching function $\sigma_k(\cdot)$ corresponding to the control component $u_k(\cdot):[t_0, t_1]\to R^1$ has at most isolated zeros on $[t_0, t_1]$, then $u_k(\cdot)$ takes on its limiting values, $u_k^{\min}$ or $u_k^{\max}$, on the entire interval. We say that the kth control component is *bang–bang*.

In this section we restrict the discussion to problems for which $f_0(x, u) \equiv 1$ so that the cost to be minimized is the interval $t_1 - t_0$. Since t often stands for time, we speak of "time-optimality." Now we consider an extremal control $u(\cdot):[t_0, t_1] \to R^m$ for a time-optimal control problem and deduce a condition that assures the bang–bang nature of the control component $u_k(\cdot)$.

Since $f_0(x, u) \equiv 1$ implies that $b_{0k} = 0$, the switching function for $u_k(\cdot)$ is given by

$$\sigma_k(t) = \sum_{i=1}^n \lambda_i(t) b_{ik}$$

or, in vector form,

$$\sigma_k(t) = \hat{\lambda}^T(t) b_k \tag{14.8}$$

where

$$\hat{\lambda} \triangleq [\lambda_1 \; \lambda_2 \cdots \lambda_n]^T, \qquad b_k \triangleq [b_{1k} \; b_{2k} \cdots b_{nk}]^T.$$

Now suppose that

$$\sigma_k(t) = 0 \qquad \forall t \in [t', t''] \tag{14.9}$$

where $t_0 \le t' < t'' \le t_1$. The adjoint equations (11.28) become

$$\dot{\hat{\lambda}}^T(t) = -\hat{\lambda}^T(t) A \tag{14.10}$$

so that the derivatives of $\sigma_k(\cdot)$ up to any given order are defined. In particular, as a consequence of (14.9), we have

$$\sigma_k(t) = \frac{d\sigma_k(t)}{dt} = \cdots = \frac{d^{n-1}\sigma_k(t)}{dt^{n-1}} = 0 \tag{14.11}$$

for all $t \in [t', t'']$. Then it follows from (14.11), together with (14.8) and (14.10), that

$$\begin{aligned}
\hat{\lambda}^T(t) b_k &= 0, \\
\hat{\lambda}^T(t) A b_k &= 0, \\
&\vdots \\
\hat{\lambda}^T(t) A^{n-1} b_k &= 0
\end{aligned} \tag{14.12}$$

for all $t \in [t', t'']$.

If the vectors $b_k, Ab_k, \ldots, A^{n-1}b_k$ are linearly independent so that the determinant of the coefficient matrix in (14.12) is not zero, then $\hat{\lambda}(t) = 0$ for all $t \in [t', t'']$. However, the condition (b) of Theorem 11.1 implies that

$$\lambda_0(t) + \hat{\lambda}^T(t)[Ax(t) + Bu(t)] = 0 \tag{14.13}$$

so that $\hat{\lambda}(t) \equiv 0$ results in $\lambda_0(t) \equiv 0$, and hence in the trivial solution $\lambda(t) \equiv 0$; but according to Theorem 11.1 there is a nontrivial solution of the adjoint equations. Thus we arrive at the following result.

Lemma 14.1. *If $u(\cdot):[t_0, t_1] \to R^m$ is an extremal control for the system (14.2)–(14.4) with $f_0(x,u) \equiv 1$ and if the vectors $b_k, Ab_k, \ldots, A^{n-1}b_k$ are linearly independent, then the component $u_k(\cdot)$ is bang–bang.*

14.5. The Number of Switches

Provided the conditions of Lemma 14.1 are met, the component $u_k(\cdot)$ of an extremal control is bang–bang; in other words, $u_k(\cdot)$ takes on only the limiting values $u_k^{\min}$ or $u_k^{\max}$. However, thus far we do not know how many times the control component $u_k(\cdot)$ can *switch* from one limiting value to the other. The following lemma provides the answer to that question.

Lemma 14.2. *Consider the system (14.2)–(14.4) with $f_0(x,u) \equiv 1$. If (a) the eigenvalues of the matrix A are real, and (b) the component $u_k(\cdot)$ of the extremal control $u(\cdot):[t_0, t_1] \to R^m$ is bang–bang, then there can be at most $n-1$ switches in the value of $u_k(\cdot)$.*

Proof. If all eigenvalues of A are real, then each component of $\lambda_j(\cdot)$, $j = 1, 2, \ldots, n$, of a solution $\hat{\lambda}(\cdot):[t_0, t_1] \to R^n$ of (14.10) has values of the form

$$\phi_1(t)e^{\alpha_1 t} + \phi_2(t)e^{\alpha_2 t} + \cdots + \phi_\gamma(t)e^{\alpha_\gamma t}, \tag{14.14}$$

where the $\phi_i(t)$, $i = 1, 2, \ldots, \gamma$, are polynomials in t and the α_i are the *distinct* eigenvalues of the matrix $-A$ (for instance, see Ref. 14.1).

The switching function $\sigma_k(\cdot)$, defined by (14.8), is a linear combination of the adjoint vector components $\lambda_j(\cdot)$ and hence is of the same form as (14.14). Thus in order to show that the number of zeros of $\sigma_k(\cdot)$, and hence the number of switches of $u_k(\cdot)$, cannot exceed $n-1$, it suffices to demonstrate that a function whose value is of the form (14.14) has at most $n-1$ zeros on $[t_0, t_1]$.

We prove first that the number of zeros of a function given by (14.14) cannot be greater than $d_1 + d_2 + \cdots + d_\gamma + \gamma - 1$, where d_i denotes the degree of the polynomial $\phi_i(t)$. This assertion is clearly valid if $\gamma = 1$ since then (14.14) is $\phi_1(t)e^{\alpha_1 t}$ and so has at most d_1 zeros. Now we employ an argument by induction to show that the assertion is valid for $\gamma > 1$; that is, we prove that the assertion holds for (14.14) with γ terms if it holds for (14.14) with $\gamma - 1$ terms.

Multiply the expression (14.14) by $e^{-\alpha_\gamma t}$, yielding

$$\phi_1(t)e^{(\alpha_1 - \alpha_\gamma)t} + \cdots + \phi_{\gamma-1}(t)e^{(\alpha_{\gamma-1} - \alpha_\gamma)t} + \phi_\gamma(t). \qquad (14.15)$$

Clearly the function given by (14.15) has the same number of zeros as does the one given by (14.14). Suppose now that the assertion is false, that is, that the function given by (14.14), and hence the one given by (14.15), has at least $d_1 + d_2 + \cdots + d_\gamma + \gamma$ zeros on $[t_0, t_1]$. Between two successive zeros of a continuously differentiable function there is at least one zero of its derivative. Consequently the $(d_\gamma + 1)$-th derivative of (14.15) has at least $(d_1 + d_2 + \cdots + d_\gamma + \gamma) - (d_\gamma + 1)$ zeros. However the $(d_\gamma + 1)$-th derivative of (14.15) is of the form

$$\bar{\phi}_1(t)e^{(\alpha_1 - \alpha_\gamma)t} + \cdots + \bar{\phi}_{\gamma-1}(t)e^{(\alpha_{\gamma-1} - \alpha_\gamma)} \qquad (14.16)$$

where $\bar{\phi}_i(t)$ is a polynomial of degree d_i and the $\alpha_i - \alpha_\gamma$, $i = 1, 2, \ldots, \gamma - 1$, are distinct numbers. In other words, (14.16) is of the same form as (14.14), except for having $\gamma - 1$ terms rather than γ terms. Now if the assertion is valid for a function of the form (14.14) with $\gamma - 1$ terms, and hence for (14.16), then (14.16) has at most $d_1 + d_2 + \cdots + d_{\gamma-1} + (\gamma - 1) - 1$ zeros. Thus supposing that the assertion holds for a function of the form (14.14) with $\gamma - 1$ terms but not with γ terms leads to a contradiction. Since the assertion is valid for $\gamma = 1$ it is also valid for $\gamma > 1$.

If m_i denotes the multiplicity of the eigenvalue α_i, then

$$\sum_{i=1}^{\gamma} m_i = n.$$

Furthermore, $d_i < m_i$ so that

$$d_1 + d_2 + \cdots + d_\gamma + \gamma - 1 = (d_1 + 1) + (d_2 + 1) + \cdots + (d_\gamma + 1) - 1$$
$$\leq m_1 + m_2 + \cdots + m_\gamma - 1$$
$$\leq n - 1.$$

This concludes the proof of Lemma 14.2. □

Before continuing the discussion of some special systems let us consider some examples in order to illustrate the results obtained so far.

14.6. A One-Dimensional Time-Optimal Regulator

Consider a system, such as a unit mass in rectilinear motion subjected to a bounded force, with the state equations

$$\dot{x}_1(t) = x_2(t), \quad \dot{x}_2(t) = u(t), \tag{14.17}$$

and the control constrained by

$$|u(t)| \leq 1. \tag{14.18}$$

It is desired to transfer the state from a given initial one, x^0, to $x^1 = 0$ and to accomplish this in minimum "time." Thus we have a problem of the type discussed in Section 14.2 with $f_0(x, u) \equiv 1$.

To deduce an extremal control $u(\cdot):[t_0, t_1] \to R^1$ we invoke the necessary conditions embodied in the maximum principle, Theorem 11.1. The function $H(\cdot)$ is now given by

$$H(\lambda, y, u) = \lambda_0 + \lambda_1 x_2 + \lambda_2 u$$

so that the adjoint equations for $i = 1, 2$ are

$$\dot{\lambda}_1(t) = 0, \quad \dot{\lambda}_2(t) = -\lambda_1(t)$$

whence one obtains

$$\lambda_1(t) = \lambda_1(t_1), \quad \lambda_2(t) = \lambda_2(t_1) + \lambda_1(t_1)(t_1 - t).$$

Thus we have

$$H[\lambda(t), y(t), u] = \lambda_0(t) + c_1 x_2(t) + (c_2 - c_1 t) u \tag{14.19}$$

where $x(\cdot): [t_0, t_1] \to R^2$ is the solution generated by $u(\cdot)$, and

$$c_1 \triangleq \lambda_1(t_1) = \text{constant}, \qquad c_2 \triangleq \lambda_1(t_1)t_1 + \lambda_2(t_1) = \text{constant}.$$

The switching function $\sigma(\cdot)$ for $u(\cdot)$ is then given by

$$\sigma(t) = \lambda_2(t) = c_2 - c_1 t. \tag{14.20}$$

Since $\sigma(\cdot)$ is a linear function it has at most one zero unless $c_1 = c_2 = 0$. If $c_1 = c_2 = 0$, then $\lambda_1(t) = \lambda_2(t) \equiv 0$ and, as a consequence of the condition (b) of Theorem 11.1, $\lambda_0(t) \equiv 0$ as well. Since there must exist a nonzero solution of the adjoint equations, not both c_1 and c_2 are zero. Thus the extremal control $u(\cdot)$ is bang–bang and has at most one switch. It is readily established that these conclusions conform to Lemmas 14.1 and 14.2.

In order to apply Lemma 14.1 we require the determinant of the coefficient matrix of equations (14.12); for the problem at hand it is simply equal to one. Hence the conditions of the lemma are met; an extremal control is bang–bang.

Since the eigenvalues of the coefficient matrix A for the state equations (14.17) are zero, they are real, and so the bang–bang extremal control $u(\cdot)$ has at most $n - 1 = 2 - 1 = 1$ switch.

Having established that an extremal control is bang–bang with at most one switch, we turn to a discussion of the trajectories generated by an extremal control. The integration of the state equations (14.17) yields

$$\begin{aligned} x_2^2(t) &= 2x_1(t) + k_+ & \text{for } u(t) \equiv 1, \\ x_2^2(t) &= -2x_1(t) + k_- & \text{for } u(t) \equiv -1, \end{aligned} \tag{14.21}$$

where k_+ and k_- are constants; thus equations (14.21) define two one-parameter families of parabolic trajectories (see Figure 14.1).

In view of the second state equation we have

$$\begin{aligned} \dot{x}_2(t) &> 0 & \text{if } u(t) = 1, \\ \dot{x}_2(t) &< 0 & \text{if } u(t) = -1, \end{aligned}$$

defining thereby the direction of increasing "time" on a trajectory; this direction is indicated by the arrows in Figure 14.1.

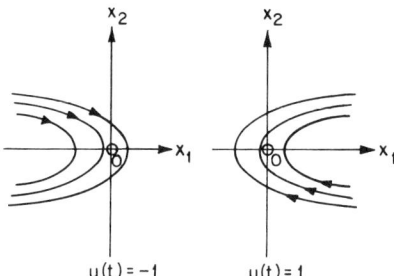

Figure 14.1. Trajectories.

Since there can be at most one switch in the value of an extremal control, a corresponding trajectory is composed of no more than two parabolic arcs which belong to two of the intersecting parabolas defined by (14.21). Furthermore, since the trajectory generated by an extremal control must terminate at $x^1 = 0$, the terminal arc—that is, the second one if there are two—belongs to one of the half parabolas defined by

$$x_2^2(t) = 2x_1(t) \quad [\dot{x}_2(t) > 0],$$
$$x_2^2(t) = -2x_1(t) \quad [\dot{x}_2(t) < 0].$$

These half parabolas, denoted by AO and BO, respectively, in Figure 14.2, compose the *switching curve* in the state space; that is, AOB is the locus of states where an extremal control switches. The switching curve divides the state space into two open sets. The trajectories which emanate from the

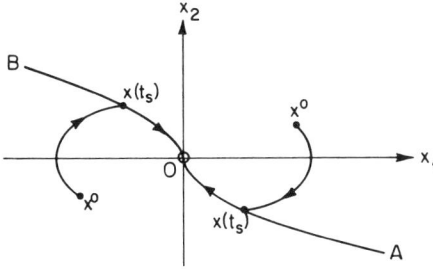

Figure 14.2. The switching curve.

states lying below AOB correspond to the extremal control sequence

$$\begin{aligned} u(t) &= 1 && \text{for } t \in [t_0, t_s), \\ u(t) &= -1 && \text{for } t \in [t_s, t_1] \end{aligned} \tag{14.22}$$

with $x(t_s) \in AOB$, and the trajectories starting at the states in the region above AOB are generated by the extremal control

$$\begin{aligned} u(t) &= -1 && \text{for } t \in [t_0, t_s), \\ u(t) &= 1 && \text{for } t \in [t_s, t_1] \end{aligned} \tag{14.23}$$

with $x(t_s) \in AOB$. If the initial state, x^0, belongs to the switching curve, there is no switch in the value of $u(\cdot)$; that is,

$$u(t) = 1 \quad \text{for } t \in [t_0, t_1]$$

or

$$u(t) = -1 \quad \text{for } t \in [t_0, t_1].$$

The following two conclusions are obvious. Firstly, given the initial state x^0, the extremal control is unique. Secondly, for every initial state in $R^2 \setminus \{0\}$ there exists an extremal control. Nonetheless we cannot yet assert that the extremal control is optimal; in fact, it is optimal (Exercise 15.10).

Accepting for the moment the fact that the extremal control for a given initial state is indeed optimal, it is of interest to deduce the optimal cost function $V^*(\cdot): R^2 \to R^1$, where

$$V^*(x^0) = \int_{t_0}^{t_1^*} dt = t_1^* - t_0.$$

This is readily done by integrating the state equations with the control sequences (14.22) and (14.23), respectively (Exercise 14.1). This yields

$$V^*(x^0) = \begin{cases} x_2^0 + \left[4x_1^0 + 2(x_2^0)^2 \right]^{1/2} & \text{for } x_1^0 > -\frac{1}{2} x_2^0 |x_2^0|, \\ -x_2^0 + \left[-4x_1^0 + 2(x_2^0)^2 \right]^{1/2} & \text{for } x_1^0 < -\frac{1}{2} x_2^0 |x_2^0|, \\ |x_2^0| & \text{for } x_1^0 = -\frac{1}{2} x_2^0 |x_2^0|. \end{cases} \tag{14.24}$$

Chap. 14 • Special Systems

Now one can construct the family of optimal isocost surfaces $\{S(C)\}$, where

$$S(C) \triangleq \{x^0 \in R^2 | V^*(x^0) = C\};$$

that is, the *minimum* transfer "time," $t_1^* - t_0$, is the same for every initial state belonging to a given isocost surface. In view of (14.24), an isocost surface consists of two parabolic arcs which intersect on the switching curve; namely, if $x^0 \in S(C)$ then

$$x_1^0 = \begin{cases} -\frac{1}{2}(x_2^0)^2 + \frac{1}{4}(C - x_2^0)^2 & \text{for } x_1^0 > -\frac{1}{2}x_2^0|x_2^0|, \\ \frac{1}{2}(x_2^0)^2 - \frac{1}{4}(C + x_2^0)^2 & \text{for } x_1^0 < -\frac{1}{2}x_2^0|x_2^0|, \\ -\frac{1}{2}x_2^0 C & \text{for } x_1^0 = -\frac{1}{2}x_2^0|x_2^0|. \end{cases}$$

This is illustrated in Figure 14.3. It follows at once from (14.24) that grad $V^*(x)$ is defined for $x \notin AOB$, but it is not defined for x on the switching curve (Exercise 14.2). Thus, if $x \notin AOB$, we have

$$\left. \begin{array}{l} \dfrac{\partial V^*(x)}{\partial x_1} = \dfrac{2}{V^*(x) - x_2} \\[6pt] \dfrac{\partial V^*(x)}{\partial x_2} = \dfrac{V^*(x) + x_2}{V^*(x) - x_2} \end{array} \right\} \quad \text{for } x_1 > -\tfrac{1}{2}x_2|x_2|$$

and

$$\left. \begin{array}{l} \dfrac{\partial V^*(x)}{\partial x_1} = \dfrac{-2}{V^*(x) + x_2} \\[6pt] \dfrac{\partial V^*(x)}{\partial x_2} = \dfrac{x_2 - V^*(x)}{x_2 + V^*(x)} \end{array} \right\} \quad \text{for } x_1 < -\tfrac{1}{2}x_2|x_2|.$$

Utilizing the condition (b) of Theorem 11.1, one can then show that

$$\lambda_i(t) = -\frac{\partial V^*(x^*(t))}{\partial x_i}, \quad i = 1, 2, \tag{14.25}$$

provided $x^*(t) \notin AOB$; this is left as an exercise for the reader (Exercise 14.3).

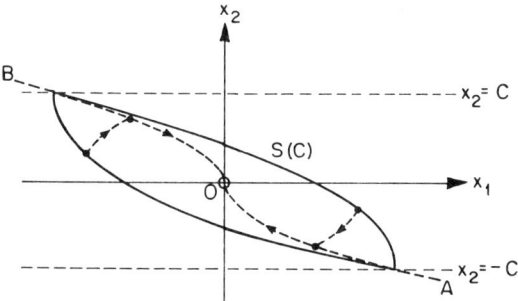

Figure 14.3. An isocost surface.

If the control $u^*(\cdot):[t_0, t_1^*] \to R^1$ is optimal at x^0 and generates the solution $x^*(\cdot):[t_0, t_1^*] \to R^2$, then the condition (b) is

$$\lambda_0(t) + \lambda_1(t)x_2^*(t) + \sigma(t)u^*(t) = 0. \tag{14.26}$$

Since $x_2^*(t_1^*) = 0$, it follows from (14.26) that $\lambda_0(t) \equiv 0$ results in $\sigma(t_1^*) = 0$. However, as shown earlier, the switching function has at most one zero on $[t_0, t_1^*]$. Consequently, if $x^0 \notin AOB$ so that there is a switch, $\lambda_0(t) \equiv \text{const} \neq 0$. On the other hand, if the initial state belongs to the switching curve so that there is no switch, one can no longer show that $\lambda_0(t) \neq 0$. Indeed then the adjoint equations possess a two-parameter family of solutions for which the maximum principle is satisfied, and one of these solutions is such that $\lambda_0(t) \equiv 0$ (Exercise 14.4).

14.7. A Three-Dimensional Time-Optimal Regulator

Let us now consider the three-dimensional version of the example discussed in Section 14.6, that is, a system with the state equations

$$\begin{aligned} \dot{x}_i(t) &= x_{i+3}(t), \\ \dot{x}_{i+3}(t) &= u_i(t), \quad i = 1, 2, 3, \end{aligned} \tag{14.27}$$

and the control constrained by

$$|u_i(t)| \leq 1, \quad i = 1, 2, 3. \tag{14.28}$$

It is again desired to transfer the state from a given initial one, x^0, to $x^1 = 0$ and to do so in minimum "time."

Now the function $H(\cdot)$ is given by

$$H(\lambda, y, u) = \lambda_0 + \sum_{i=1}^{3} \lambda_i x_{i+3} + \sum_{i=1}^{3} \lambda_{i+3} u_i$$

so that the adjoint equations for $i = 1, 2, 3$ are

$$\dot{\lambda}_i(t) = 0, \qquad \dot{\lambda}_{i+3}(t) = -\lambda_i(t),$$

whence we obtain

$$\lambda_i(t) = \lambda_i(t_1), \qquad \lambda_{i+3}(t) = \lambda_{i+3}(t_1) + \lambda_i(t_1)(t_1 - t),$$

so that

$$H[\lambda(t), y(t), u] = \lambda_0(t) + \sum_{i=1}^{3} c_i x_{i+3}(t) + \sum_{i=1}^{3} (c_{i+3} - c_i t) u_i \quad (14.29)$$

where

$$c_i \stackrel{\Delta}{=} \lambda_i(t_1) = \text{constant},$$

$$c_{i+3} \stackrel{\Delta}{=} \lambda_i(t_1) t_1 + \lambda_{i+3}(t_1) = \text{constant},$$

and $x(\cdot):[t_0, t_1] \to R^6$ is the solution generated by the extremal control $u(\cdot):[t_0, t_1] \to R^3$.

The switching function for $u_i(\cdot)$ is $\sigma_i(\cdot)$ given by

$$\sigma_i(t) = c_{i+3} - c_i t.$$

Since $\sigma_i(\cdot)$ is a linear function, it can have at most one zero unless $c_i = c_{i+3} = 0$. However, by condition (b) of Theorem 11.1, we can rule out

$$\sigma_i(t) \equiv 0, \quad i = 1, 2, 3$$

since in that event $\lambda_i(t) \equiv 0$, $i = 0, 1, \ldots, 6$. Consequently at least one of the switching functions is not identically zero so that the corresponding control component is bang–bang. One cannot show that *each* of the switching functions can have at most one zero, that is, that all three control components are bang–bang. Indeed the condition in Lemma 14.1, assuring that $u_k(\cdot)$ is bang–bang, is not satisfied.

For the moment consider the system to be composed of three *independent* (uncoupled) subsystems, each of which is governed by an equation of the form

$$\ddot{x}_i(t) = u_i(t)$$

with

$$|u_i(t)| \leq 1,$$

and for which the state $[x_i(t_0) \; \dot{x}_i(t_0)]^T$ is to be transferred from a given one to zero with the minimum value of the transfer time $t_1^i - t_0$. Each subsystem is one of the type discussed in Section 14.6. Let $u^{i*}(\cdot):[t_0, t_1^{i*}] \to R^1$ denote the optimal control at the given initial state for the ith subsystem, and let τ^i be the minimum transfer time, that is, $\tau^i = t_1^{i*} - t_0$.

Of course the system is not made up of three *independent* subsystems. Thus the minimum value of the transfer time of the system cannot be smaller than the largest of the three minimum values of the transfer times obtained by treating each subsystem as being independent. Hence, for example, if

$$\tau^1 > \tau^2 \quad \text{and} \quad \tau^1 > \tau^3,$$

then the minimum value of the system's transfer time is

$$t_1^* - t_0 = \tau^1.$$

The optimal control $u^*(\cdot)$ is no longer unique; namely, $u_1^*(\cdot) = u^{1*}(\cdot)$ is bang–bang, while $u_2^*(\cdot)$ and $u_3^*(\cdot)$ can be *any* admissible functions resulting in the transfer of $[x_2^*(t_0) \; \dot{x}_2^*(t_0)]^T$ and $[x_3^*(t_0) \; \dot{x}_3^*(t_0)]^T$ to $[0 \; 0]^T$ in the interval τ^1. For instance, for $i = 2, 3$ we may choose

$$u_i^*(t) = u^{i*}(t) \qquad \text{for } t \in [t_0, t_1^{i*}),$$
$$u_i^*(t) = 0 \qquad \text{for } t \in [t_1^{i*}, t_1^*].$$

The determination of the time-optimal control is straightforward; it involves merely the determination of the time-optimal control for each subsystem considered as an independent one.

14.8. Singular Control

Now we turn to systems for which the functions $f_i(\cdot)$ depend in a linear fashion on some of the control components. Let

$$u \triangleq \begin{bmatrix} u^n \\ -- \\ u^l \end{bmatrix}$$

where

$$u^n \triangleq [u_1 \; u_2 \cdots u_r]^T,$$

$$u^l \triangleq [u_{r+1} \; u_{r+2} \cdots u_m]^T.$$

Suppose that

$$f_i(x, u) = a_i(x, u^n) + \sum_{j=1}^{m-r} b_{ij}(x, u^n) u_{r+j}, \qquad i=0,1,\ldots,n,$$

where the $a_i(\cdot)$ and $b_{ij}(\cdot)$ are given functions from $R^n \times R^r$ into R^1, and of course such that the earlier continuity assumptions on $f_i(\cdot)$ are satisfied. In other words, given the control $u(\cdot):[t_0, t_1] \to R^m$, the state equations are

$$\dot{x}_i(t) = a_i[x(t), u^n(t)] + \sum_{j=1}^{m-r} b_{ij}[x(t), u^n(t)] u_{r+j}(t), \qquad i=1,2,\ldots,n \tag{14.30}$$

and the cost is

$$\int_{t_0}^{t_1} \left\{ a_0[x(t), u^n(t)] + \sum_{j=1}^{m-r} b_{0j}[x(t), u^n(t)] u_{r+j}(t) \right\} dt. \tag{14.31}$$

We suppose further that the prescribed control constraint set U is such that u^l is constrained solely by

$$u_i^{\min} \leq u_i \leq u_i^{\max}, \qquad i=r+1, r+2,\ldots, m, \tag{14.32}$$

where the $u_i^{\min}$ and $u_i^{\max}$ are given constants.

As before we are interested in the extremal controls. The function $H(\cdot)$ is now given by

$$H(\lambda, y, u) = \sum_{i=0}^{n} \lambda_i a_i(x, u^n) + \sum_{i=0}^{n} \sum_{j=1}^{m-r} \lambda_i b_{ij}(x, u^n) u_{r+j}.$$

Thus, if $u(\cdot): [t_0, t_1] \to R^m$ is an extremal control that generates the solution $x(\cdot): [t_0, t_1] \to R^n$ of the state equations and $\lambda(\cdot): [t_0, t_1] \to R^{n+1}$ of the adjoint equations, then $\sigma_j(\cdot): [t_0, t_1] \to R^1$ given by

$$\sigma_j(t) = \sum_{i=0}^{n} \lambda_i(t) b_{ij}[x(t), u^n(t)] \tag{14.33}$$

is the *switching function* for the control component $u_{r+j}(\cdot)$. In other words,

$$u_{r+j}(t) = u_{r+j}^{\min} \quad \text{if } \sigma_j(t) < 0,$$
$$u_{r+j}(t) = u_{r+j}^{\max} \quad \text{if } \sigma_j(t) > 0.$$

In particular cases it may be possible to show that $\sigma_j(\cdot)$ can have at most isolated zeros so that the corresponding control component is bang–bang; for instance see the example treated in Section 14.9. On the other hand there may exist an extremal control such that $\sigma_j(t) \equiv 0$ on a subinterval of $[t_0, t_1]$; the corresponding extremal control component, $u_j(\cdot)$, is termed *singular*; for instance see the example of Section 14.10. A full treatment of singular extremal control is beyond the scope of our discussion; a rather complete treatment may be found in Ref. 14.2.

14.9. The Maximum Range of a Thrust-Limited Rocket

Let us recall the system discussed in Section 13.5, namely, a rocket for which the state equations are

$$\dot{x}_1(t) = x_3(t),$$
$$\dot{x}_2(t) = x_4(t),$$
$$\dot{x}_3(t) = \frac{c}{x_5(t)} u_1(t) u_3(t), \tag{14.34}$$
$$\dot{x}_4(t) = \frac{c}{x_5(t)} u_2(t) u_3(t) - g,$$
$$\dot{x}_5(t) = -u_3(t).$$

Chap. 14 • Special Systems

The control constraint set U is defined by

$$u_1^2 + u_2^2 = 1, \qquad 0 \leq u_3 \leq u_3^{\max}. \tag{14.35}$$

It is desired to choose the thrust direction and magnitude so as to transfer a rocket of given initial mass, position, and velocity to a given altitude while expending a prescribed amount of fuel *and* to attain the maximum range. Consequently the initial state is given, say x^0, while the target set is

$$\theta^1 = \{x \in R^5 | x_2 - x_2^1 = 0, \, x_5 - x_5^1 = 0\} \tag{14.36}$$

where x_2^1 and x_5^1 are given constants with $0 < x_5^1 < x_5^0$. The cost to be *minimized* is

$$-\int_{t_0}^{t_1} x_3(t) \, dt, \tag{14.37}$$

with the minus sign to account for maximization.

Thus we have a problem of the type introduced in Section 14.8, with

$$u^n = [u_1 \; u_2]^T, \qquad u^l = u_3.$$

In order to utilize the maximum principle we need the function $H(\cdot)$; for our problem it is given by

$$H(\lambda, y, u) = -\lambda_0 x_3 + \lambda_1 x_3 + \lambda_2 x_4 - \lambda_4 g + \left[\frac{c}{x_5} (\lambda_3 u_1 + \lambda_4 u_2) - \lambda_5 \right] u_3 \tag{14.38}$$

so that the adjoint equations become

$$\begin{aligned}
\dot{\lambda}_0(t) &= 0, \\
\dot{\lambda}_1(t) &= 0, \\
\dot{\lambda}_2(t) &= 0, \\
\dot{\lambda}_3(t) &= \lambda_0(t) - \lambda_1(t), \\
\dot{\lambda}_4(t) &= -\lambda_2(t), \\
\dot{\lambda}_5(t) &= \frac{c}{x_5^2(t)} \left[\lambda_3(t) u_1(t) + \lambda_4(t) u_2(t) \right] u_3(t),
\end{aligned} \tag{14.39}$$

whence we obtain

$$\begin{aligned}\lambda_1(t)&=\lambda_1(t_1),\\ \lambda_2(t)&=\lambda_2(t_1),\\ \lambda_3(t)&=[\lambda_1(t_1)-\lambda_0](t_1-t)+\lambda_3(t_1),\\ \lambda_4(t)&=\lambda_2(t_1)(t_1-t)+\lambda_4(t_1),\end{aligned} \quad (14.40)$$

where, by the condition (c) of Theorem 11.1,

$$\lambda_0(t)\equiv\text{const}\stackrel{\triangle}{=}\lambda_0\le 0.$$

The terminal transversality condition (11.38)–(11.39), together with (14.36), implies that

$$\lambda_1(t_1)=\lambda_3(t_1)=\lambda_4(t_1)=0$$

so that (14.40) becomes

$$\begin{aligned}\lambda_1(t)&=0,\\ \lambda_2(t)&=\lambda_2(t_1),\\ \lambda_3(t)&=-\lambda_0(t_1-t),\\ \lambda_4(t)&=\lambda_2(t_1)(t_1-t).\end{aligned} \quad (14.41)$$

To employ the condition (a) of Theorem 11.1 we require that part of the function $H(\cdot)$ which depends on the control, namely,

$$\left\{\frac{c}{x_5(t)}[\lambda_3(t)u_1+\lambda_4(t)u_2]-\lambda_5(t)\right\}u_3. \quad (14.42)$$

Since $u_3(t)\ge 0$ and $x_5^1>0$, we have $x_5(t)>0$. Also, $c>0$. Thus, provided $u_3=u_3(t)\ne 0$, it follows from the condition (a), together with the constraint (14.35), that

$$u_i(t)=\frac{\lambda_{i+2}(t)}{[\lambda_3^2(t)+\lambda_4^2(t)]^{1/2}}, \quad i=1,2,$$

so that, in view of (14.41),

$$u_1(t) = \frac{-\lambda_0}{[\lambda_0^2 + \lambda_2^2(t_1)]^{1/2}}, \qquad u_2(t) = \frac{\lambda_2(t_1)}{[\lambda_0^2 + \lambda_2^2(t_1)]^{1/2}}. \qquad (14.43)$$

In other words, the thrust direction—when it is defined—is constant.

We turn next to a consideration of the thrust magnitude. Upon substitution of (14.41) in (14.42) the switching function of $u_3(\cdot)$, $\sigma(\cdot)$, is given by

$$\sigma(t) = \frac{c}{x_5(t)} [\lambda_0^2 + \lambda_2^2(t_1)]^{1/2} (t_1 - t) - \lambda_5(t). \qquad (14.44)$$

In order to investigate the behavior of $\sigma(\cdot)$ we form its derivative; upon use of the last of the adjoint equations (14.39) we obtain

$$\dot{\sigma}(t) = -\frac{c}{x_5(t)} [\lambda_0^2 + \lambda_2^2(t_1)]^{1/2}. \qquad (14.45)$$

As noted earlier, $c > 0$ and $x_5(t) > 0$. Thus, if $\dot{\sigma}(t) = 0$, it follows from (14.45) that $\lambda_0 = \lambda_2(t_1) = 0$, and then from (14.41) that $\lambda_i(t) \equiv 0$ for $i = 0, 1, \ldots, 4$. However in that event the condition (b) of Theorem 11.1, together with $u_3(t) \neq 0$, implies that $\lambda_5(t) \equiv 0$ as well, resulting in a zero solution of the adjoint equations. This is in conflict with Theorem 11.1, and so we conclude that

$$\dot{\sigma}(t) \neq 0 \qquad \forall t \in [t_0, t_1].$$

Then $c > 0$ and $x_5(t) > 0$ imply that

$$\dot{\sigma}(t) < 0 \qquad \forall t \in [t_0, t_1].$$

Consequently the extremal control is not singular and, in particular, the mass flow rate $u_3(\cdot)$ is bang-bang with the switching sequence

$$\begin{aligned} u_3(t) &= u_3^{\max} & \text{for } t \in [t_0, t_s), \\ u_3(t) &= 0 & \text{for } t \in [t_s, t_1]. \end{aligned}$$

The time interval during which the thrust is on, the so-called *burning time* $t_s - t_0$, can be computed by integrating the fifth of the state equations

(14.34); namely,

$$t_s - t_0 = \frac{x_5^0 - x_5^1}{u_3^{max}}.$$

We can conclude readily that $\lambda_0 = 0$ cannot correspond to an extremal control that is optimal, except in the limiting case of just being able to attain x_2^1. If $\lambda_0 = 0$ then $u_1(t) = 0$ for $u_3(t) \neq 0$, so that $\dot{x}_3(t) \equiv 0$ and hence $x_3(t) \equiv x_3^0$; in other words, the control—here the rocket thrust—has no effect on the cost—here the range. For $\lambda_0 \neq 0$, the condition (b) of Theorem 11.1 at $t = t_1 > t_s$, namely,

$$-\lambda_0 x_3(t_1) + \lambda_2(t_1) x_4(t_1) = 0,$$

together with (14.43), can be solved for the thrust-direction angle (see Figure 14.4)

$$\psi(t) = \tan^{-1}\left[\frac{u_2(t)}{u_1(t)}\right] \equiv \tan^{-1}\left[-\frac{x_3(t_1)}{x_4(t_1)}\right]. \qquad (14.46)$$

In other words, when the thrust is on its direction is constant. On integration of the third and fourth of the state equations (14.34) over $[t_s, t_1]$ we have

$$x_3(t_1) = x_3(t_s),$$

$$x_4(t_1) = -\{x_4^2(t_s) - 2g[x_2(t_1) - x_2(t_s)]\}^{1/2},$$

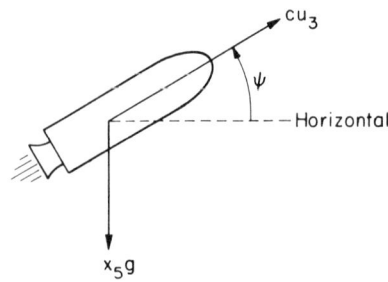

Figure 14.4. Forces acting on a rocket.

so that

$$\psi(t) \equiv \tan^{-1} \frac{x_3(t_s)}{\{x_4^2(t_s) - 2g[x_2^1 - x_2(t_s)]\}^{1/2}} \triangleq \psi. \qquad (14.47)$$

Finally the integration of the state equations over $[t_0, t_s]$ yields

$$x_3(t_s) = x_3^0 + c \ln\left(\frac{x_5^0}{x_5^1}\right) \cos\psi,$$

$$x_4(t_s) = x_4^0 + c \ln\left(\frac{x_5^0}{x_5^1}\right) \sin\psi - g(t_s - t_0), \qquad (14.48)$$

$$x_2(t_s) = x_2^0 + c \sin\psi \int_{t_0}^{t_s} \left(\ln\frac{x_5^0}{x_5^0 - u_3^{\max} t}\right) dt - \frac{1}{2} g(t_s - t_0)^2.$$

The value of ψ is obtained by iteration, employing (14.47) in conjunction with (14.48).

14.10. The Maximum Range of a Rocket in Horizontal Flight

Now we take up the problem of maximizing the distance traveled by a rocket along a horizontal rectilinear path in the atmosphere. Earlier discussions of this problem can be found in Refs. 14.3 and 14.4.

Let t be the time, x_1 be the horizontal velocity component, x_2 be the rocket mass, u be the mass flow rate, c be the effective exhaust speed = constant > 0, and g be the magnitude of gravitational acceleration.

We make the following assumptions (see Fig. 14.5):

(1) The gravitational acceleration is constant.
(2) The aerodynamic lift, L, balances the weight, $x_2 g$.
(3) The aerodynamic drag, D, depends on the velocity and on the lift; in particular, we let (see Ref. 14.5)

$$D = \alpha x_1^2 + \beta \frac{L^2}{g^2 x_1^2},$$

where α and β are positive constants.

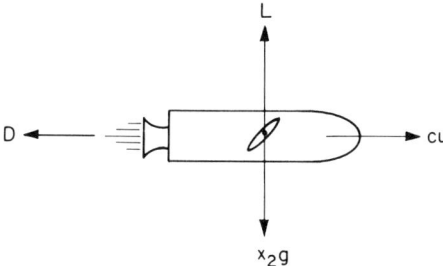

Figure 14.5. Forces acting on a rocket.

Assumption (2) combined with assumption (3) leads to

$$D = \alpha x_1^2 + \beta \frac{x_2^2}{x_1^2} \triangleq \mathcal{D}(x_1, x_2). \tag{14.49}$$

The state equations are

$$\dot{x}_1(t) = \frac{cu(t) - \mathcal{D}[x_1(t), x_2(t)]}{x_2(t)},$$
$$\dot{x}_2(t) = -u(t). \tag{14.50}$$

The mass flow rate is bounded so that the control constraint set, U, is defined by

$$0 \leq u \leq u^{\max} \tag{14.51}$$

where $u^{\max} > 0$ is a prescribed constant.

It is desired to transfer the rocket between given initial and terminal values of the velocity and mass while maximizing the range. In other words, the initial state, x^0, and the terminal state, x^1, are specified with $x_1^0 > 0$, $x_1^1 > 0$, and $0 < x_2^1 < x_2^0$. The cost to be minimized is

$$-\int_{t_0}^{t_1} x_1(t)\, dt. \tag{14.52}$$

Thus we have a problem of the kind discussed in Section 14.8.

In view of (14.50) and (14.52) we have

$$H(\lambda, y, u) = -\lambda_0 x_1 - \lambda_1 \frac{\mathcal{D}(x_1, x_2)}{x_2} + \left(\lambda_1 \frac{c}{x_2} - \lambda_2\right) u \tag{14.53}$$

so that the adjoint equations become

$$\dot{\lambda}_0(t)=0,$$

$$\dot{\lambda}_1(t)=\lambda_0(t)+\frac{\lambda_1(t)}{x_2(t)}\frac{\partial \mathcal{D}[x_1(t),x_2(t)]}{\partial x_1}, \qquad (14.54)$$

$$\dot{\lambda}_2(t)=\lambda_1(t)\left\{\frac{cu(t)-\mathcal{D}[x_1(t),x_2(t)]}{x_2^2(t)}+\frac{1}{x_2(t)}\frac{\partial \mathcal{D}[x_1(t),x_2(t)]}{\partial x_2}\right\},$$

where $u(\cdot):[t_0,t_1]\to R^1$ is an extremal control that generates the solution $x(\cdot):[t_0,t_1]\to R^2$.

The switching function for $u(\cdot)$, $\sigma(\cdot)$, is defined by

$$\sigma(t)=\lambda_1(t)\frac{c}{x_2(t)}-\lambda_2(t) \qquad (14.55)$$

so that

$$\begin{aligned} u(t)&=u^{\max} && \text{if } \sigma(t)>0, \\ u(t)&=0 && \text{if } \sigma(t)<0. \end{aligned} \qquad (14.56)$$

Next we consider the possibility of a singular control corresponding to the vanishing of $\sigma(\cdot)$ on a nonzero subinterval of $[t_0,t_1]$. To investigate this possibility we form the derivative of $\sigma(\cdot)$ for $\sigma(t)=0$, namely, $\dot{\sigma}(t)|_{\sigma(t)=0}$. As a result of the condition (b) of Theorem 11.1 we have

$$-\lambda_0(t)x_1(t)-\lambda_1(t)\frac{\mathcal{D}[x_1(t),x_2(t)]}{x_2(t)}+\sigma(t)u(t)=0 \qquad (14.57)$$

where, by the condition (c),

$$\lambda_0(t)\equiv\text{constant}\triangleq\lambda_0\le 0.$$

Upon employing (14.49), (14.50), (14.54), and (14.57) we obtain

$$\dot{\sigma}(t)|_{\sigma(t)=0}=\lambda_1(t)\mathcal{S}[x_1(t),x_2(t)] \qquad (14.58)$$

where

$$\mathcal{S}(x_1,x_2)\triangleq\beta\frac{3c+x_1}{x_1^3x_2^2}\left(\frac{\alpha}{\beta}\frac{c+x_1}{3c+x_1}x_1^4-x_2^2\right). \qquad (14.59)$$

It is easily shown that $\lambda_0 \neq 0$ if $\sigma(t)=0$ at a $t \in [t_0, t_1]$. For, in view of (14.57) and $\mathcal{D}[x_1(t), x_2(t)] > 0$, $\lambda_0 = 0$ implies that $\lambda_1(t) = 0$. Then $\sigma(t) = 0$ implies that $\lambda_2(t) = 0$ as well, resulting in the zero solution of the adjoint equations. We conclude that $\lambda_0 < 0$ if $\sigma(t) = 0$ for $t \in [t_0, t_1]$. In this connection we note that $u(t) \equiv 0$ or $u(t) \equiv u^{\max}$ if $\lambda_0 = 0$.

We are only concerned with $x_1(t) > 0$ and $x_2(t) > 0$, since this is assured by $x_1^1 > 0$ and $x_2^1 > 0$. For $x_1(t) > 0$, $x_2(t) > 0$ and $\lambda_0 < 0$ it follows from (14.57) that $\lambda_1(t) > 0$, and hence the sign of $\dot{\sigma}(t)|_{\sigma(t)=0}$ is the same as that of $\mathcal{S}[x_1(t), x_2(t)]$.

If $u(\cdot)$ is singular on a subinterval $[t', t'']$ of $[t_0, t_1]$ then $\sigma(t) = 0$ for all $t \in [t', t'']$. The locus of corresponding states is the curve $\mathcal{S}$ defined by $\mathcal{S}(x_1, x_2) = 0$; namely,

$$\mathcal{S} \triangleq \left\{ x \in R^2 \,\Big|\, x_2 = x_1^2 \left(\frac{\alpha}{\beta} \frac{c+x_1}{3c+x_1} \right)^{1/2}, x_1 > 0, x_2 > 0 \right\}. \quad (14.60)$$

This curve divides the positive quadrant into two disjoint open regions denoted by I and II, respectively, in Figure 14.6, such that

$$\begin{aligned}
\dot{\sigma}(t)|_{\sigma(t)=0} &= 0 && \text{for } x(t) \in \mathcal{S}, \\
\dot{\sigma}(t)|_{\sigma(t)=0} &< 0 && \text{for } x(t) \in \text{I}, \\
\dot{\sigma}(t)|_{\sigma(t)=0} &> 0 && \text{for } x(t) \in \text{II}.
\end{aligned} \quad (14.61)$$

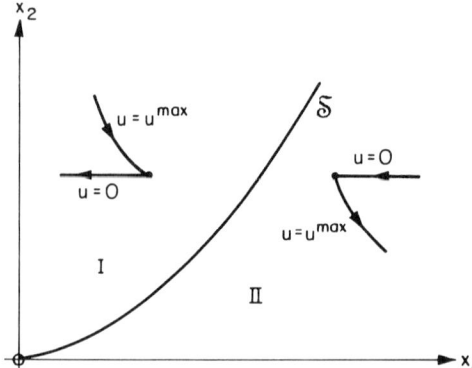

Figure 14.6. Switching sequence.

Since the switching function, $\sigma(\cdot)$, is continuous it passes through zero if it changes sign, that is, if the extremal control switches from one limiting value to the other. As a consequence of conditions (14.61) there can be at most one switch in each region. In particular, if there is a switch the sequence is

$$u^{\max} \to 0 \quad \text{in region I,}$$
$$0 \to u^{\max} \quad \text{in region II.}$$

Also, in accordance with the state equations (14.50), we have

$$\begin{aligned} \dot{x}_1(t) < 0 & \quad \text{for } u(t) = 0, \\ \dot{x}_2(t) < 0 & \quad \text{for } u(t) \in (0, u^{\max}]. \end{aligned} \qquad (14.62)$$

The corresponding situations are portrayed in Figure 14.6 where the arrows indicate the direction of increasing time.

If $u(\cdot)$ is singular on $[t', t''] \subset [t_0, t_1]$ so that $x(t) \in \mathcal{S}$ for $t \in [t', t'']$, then it follows from (14.60) that

$$x_2(t) = x_1^2(t) \left[\frac{\alpha}{\beta} \frac{c + x_1(t)}{3c + x_1(t)} \right]^{1/2} \qquad \forall t \in [t', t''].$$

This relation, together with the state equations, implies the existence of a function $k(\cdot): \mathcal{S} \to R^1$ such that

$$u(t) = k[x(t)], \qquad t \in [t', t'').$$

The computation of $k(\cdot)$ is left as an exercise for the reader (Exercise 14.6). Since an extremal control must be admissible, it must satisfy the constraint

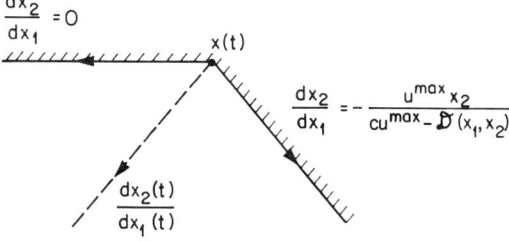

Figure 14.7. Slope bounds.

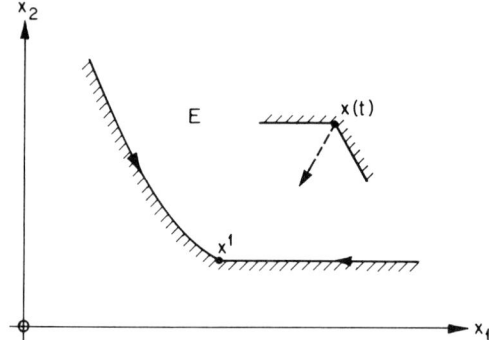

Figure 14.8. Initial state region.

(14.51). It is shown readily that the value $u(t)$, $t \in [t', t'')$, satisfies (14.51) provided the maximum thrust is sufficiently high; in particular, for $x \in \mathfrak{S}$

$$cu^{\max} \geq \mathfrak{D}(x_1, x_2) \Rightarrow k(x) \in [0, u^{\max}].$$

Now, provided the maximum thrust is sufficiently high so that a singular control exists, one can show that the extremal control is unique. Before doing so let us consider the region of initial states, E, from where the given terminal state, x^1, can be reached by means of an admissible control. In view of the state equations, the slope of a trajectory is

$$\frac{dx_2(t)}{dx_1(t)} = -\frac{x_2(t)u(t)}{cu(t) - \mathfrak{D}[x_1(t), x_2(t)]}. \tag{14.63}$$

This relation, together with the control constraint (14.51) and conditions (14.62), implies that the slope is bounded as shown in Figure 14.7. The boundary of the region E is obtained by backward integration of (14.63)

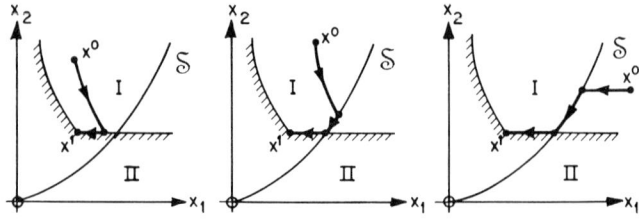

Figure 14.9. Typical trajectories generated by extremal controls.

with $u(t) \equiv 0$ and $u(t) \equiv u^{\max}$, respectively; this is illustrated in Figure 14.8. Finally, Figure 14.9 shows some typical trajectories generated by the unique extremal control for various initial states and with the prescribed terminal state $x^1 \in I$. Analogous results are obtained if $x^1 \in II$ or $x^1 \in S$.

We conclude with the reminder that, even though the extremal control for a given initial state is unique, its optimality has not been proved.

Exercises

14.1. Recall the "time-optimal" regulator problem of Section 14.6. Derive an expression for the optimal cost, $t_1^* - t_0$, as a function of the initial state, x^0. Derive an expression for the switching time, t_s.

14.2. Recall again the example discussed in Section 14.6. Show that $\operatorname{grad} V^*(x)$ is not defined for x belonging to the switching curve AOB.

14.3. For the example treated in Section 14.6 show that relations (14.25) are valid if $x^*(t) \notin AOB$.

14.4. For the problem discussed in Section 14.6 show that the maximum principle is satisfied for every solution of the adjoint equations with

$$\lambda(t_0) = \left[-\alpha_1 \left(\alpha_1/x_2^0 \right) \pm \alpha_2 \ -\alpha_2 x_2^0 \right]^T$$

where $\alpha_1 \geq 0$ and $\alpha_2 \geq 0$, $\alpha_1 + \alpha_2 = 1$, provided $x^0 \in AOB$.

14.5. For the system considered in Section 14.6 deduce candidates for the "time-optimal" control for transfer from a given initial state, x^0, to one on

$$\theta^1 = \{ x \in R^2 | x_1 = x_1^1 \}$$

and

$$\theta^1 = \{ x \in R^2 | x_2 = x_2^1 \}.$$

14.6. For the example discussed in Section 14.10 derive the function $k(\cdot)$ that defines the value of the extremal control for $x(t) \in S$. Show that $cu^{\max} \geq \mathcal{D}(x_1, x_2)$ implies that $k(x) \in [0, u^{\max}]$ for $x \in S$.

14.7. Consider the system with the state equations

$$\dot{x}_1(t) = x_2(t), \qquad \dot{x}_2(t) = -x_1(t) + u(t),$$

and the control constraint

$$|u(t)| \leq 1.$$

Discuss "time-optimal" control for the transfer from x^0 to $x^1 = 0$.

14.8. Consider the system with the state equations
$$\dot{x}_1(t)=x_2(t)+u_1(t), \qquad \dot{x}_2(t)=-x_1(t)+u_2(t),$$
subject to the control constraints
$$|u_i(t)|\le 1, \quad i=1,2.$$
Discuss "time-optimal" control for the transfer from x^0 to $x^1=0$.

14.9. Consider the system with the state equations
$$\dot{x}_i(t)=x_{i+3}(t), \qquad \dot{x}_{i+3}(t)=u_i(t), \quad i=1,2,3,$$
subject to the control constraint
$$0\le \sum_{i=1}^{3} u_i^2(t)\le 1.$$
Discuss "time-optimal" control for the transfer from x^0 to $x^1=0$.

14.10. Consider the system with the state equations
$$\dot{x}_1(t)=x_2(t)+u(t), \qquad \dot{x}_2(t)=-u(t),$$
subject to the control constraint
$$|u(t)|\le 1.$$
It is desired to transfer the state from x^0 to $x^1=0$ while minimizing the value of
$$\int_{t_0}^{t_1} x_1^2(t)\, dt.$$
Discuss extremal control.

14.11. The following state equations furnish an approximate description of the motion of an aircraft in planar flight:
$$\dot{x}_1(t)=x_2(t)u(t), \qquad \dot{x}_2(t)=F[x_1(t),x_2(t)]-u(t),$$
where $F(\cdot): R^2 \to R^1$ is a given function of class C^1. The control is constrained by
$$|u(t)|\le 1.$$
It is desired to transfer the state from x^0 to x^1 while minimizing the transfer time t_1-t_0. Discuss extremal control.

14.12. The following state equations furnish an approximate description of the rectilinear horizontal motion of a rocket:

$$\dot{x}_1(t)=x_2(t), \quad \dot{x}_2(t)=u(t),$$

with the control constrained by

$$|u(t)|\leq 1.$$

It is desired to transfer the state from x^0 to $x^1=0$ while minimizing the amount of fuel

$$\int_{t_0}^{t_1}\left[1+u^2(t)\right]^{1/2} dt.$$

Discuss extremal control.

14.13. The following state equations define the behavior of a certain hydraulic system:

$$\dot{x}_1(t)=-\alpha x_1^2(t)\operatorname{sgn} x_1(t)-\beta x_2(t)+\gamma,$$
$$\dot{x}_2(t)=\delta[x_1(t)-u(t)]$$

where α, β, γ, and δ are given positive constants. The control is constrained by

$$0\leq u(t)\leq 1.$$

It is desired to transfer the state from x^0 to x^1 while minimizing the transfer time t_1-t_0. Discuss extremal control.

14.14. Consider the system introduced in Exercise 14.13. Suppose that one wishes to transfer the state from x^0 to one in

$$\theta^1=\{x\in R^2 | x_1=x_1^1\}$$

while minimizing the value of

$$\int_{t_0}^{t_1}\left\{1+c[x_2(t)-x_2^1]^2\right\} dt$$

for given constants x_1^1, x_2^1, and $c>0$. Discuss extremal control.

15

Sufficient Conditions

15.1. Introduction

Thus far we have derived various conditions which must be satisfied if a control $u^*(\cdot):[t_0, t_1^*] \to R^m$ is optimal at x^0 for the optimal control problem posed in Chapter 10. These *necessary* conditions are embodied in the maximum principle, Theorem 11.1. Various generalizations, including "optimality on θ^0" were taken up in Chapter 13.

Necessary conditions allow one to narrow down the set of controls which can be optimal. When used constructively, necessary conditions yield controls which qualify as *candidates* for optimal ones; we term them *extremal* in order to distinguish them from those which are indeed optimal. Thus an optimal control must be extremal; however, the converse need not be true.

Now we turn to conditions whose satisfaction *assures* that a particular control is optimal at x^0, that is, *sufficient* conditions. We shall confine the discussion to problems of the type posed in Chapter 10. The various generalizations introduced in Chapter 13 can be accommodated by applying the sufficient conditions to *equivalent* systems as in Sections 13.6, 13.7, and 13.14 or to *augmented* systems as in Sections 13.4 and 13.12. We shall consider two kinds of sufficient conditions, those embodied in the so-called "field theorems" and those termed "direct" sufficient conditions.

15.2. A Field Theorem

Recall the optimal control problem posed in Chapter 10, and in particular recall the definitions of "feasible at x^0" and of "optimal at x^0." Note that a control that is optimal at x^0 is compared with *all* controls which are feasible at x^0. In the ensuing discussion we shall be more modest in that we shall require only that a control which is optimal at x^0 be compared to the feasible controls at x^0 generating trajectories which remain in a specified subset of the state space. In that sense our sufficient conditions are *local* conditions. Given a set $X \subset R^n$, $\bar{X} \cap \theta^1 \neq \emptyset$, the control $u^*(\cdot):[t_0, t_1^*] \to R^m$, generating the solution $x^*(\cdot):[t_0, t_1^*] \to R^n$, $x^*(t_0)=x^0$, *is optimal at* x^0 *with respect to X if and only if* $u^*(\cdot) \in \mathcal{U}(x^0)$, $x^*(t) \in X$ $\forall t \in [t_0, t_1^*)$ and

$$V[x^0, u^*(\cdot), x^*(\cdot)] \leq V[x^0, u(\cdot), x(\cdot)]$$

for every $u(\cdot) \in \mathcal{U}(x^0)$ generating the solution $x(\cdot):[t_0, t_1] \to R^n$, $x(t_0)=x^0$, and $x(t) \in X$ $\forall t \in [t_0, t_1)$.

Then we have the following theorem.

Theorem 15.1. *The control* $u^*(\cdot):[t_0, t_1^*] \to R^m$, *generating the solution* $x^*(\cdot):[t_0, t_1^*] \to R^n$, $x^*(t_0)=x^0$, *is optimal at* x^0 *with respect to* X (*open*) $\subset R^n$ *if there exists a function* $\mathcal{V}(\cdot): X \to R^1$ *of class* C^1 *such that*

(i) *if* $u(\cdot) \in \mathcal{U}(x^0)$ *generates the solution* $x(\cdot):[t_0, t_1] \to R^n$, $x(t_0)=x^0$, $x(t_1) \in \theta^1$ *and* $x(t) \in X$ *for all* $t \in [t_0, t_1)$, *then*

$$\lim_{t \to t_1} \mathcal{V}[x(t)] \leq \lim_{t \to t_1^*} \mathcal{V}[x^*(t)] = 0,$$

(ii) $f_0[x^*(t), u^*(t)] + \text{grad}^T \mathcal{V}[x^*(t)] f[x^*(t), u^*(t)] = 0$ $\quad \forall t \in [t_0, t_1^*)$,

(iii) $f_0(x, u) + \text{grad}^T \mathcal{V}(x) f(x, u) \geq 0$ $\quad \forall x \in X, \forall u \in U$.

Proof. Let $u(\cdot) \in \mathcal{U}(x^0)$ generate the solution $x(\cdot):[t_0, t_1] \to R^n$, $x(t_0)=x^0$, $x(t_1) \in \theta^1$ and $x(t) \in X$ for all $t \in [t_0, t_1)$. Define

$$I \triangleq \lim_{t \to t_1} \int_{t_0}^{t} \text{grad}^T \mathcal{V}[x(t)] f[x(t), u(t)] \, dt. \tag{15.1}$$

In view of the state equations (10.2) and of the condition (i) of the theorem we have

$$I \leq -\mathcal{V}(x^0). \tag{15.2}$$

Then, as a consequence of the conditions (i) and (ii) we obtain

$$\int_{t_0}^{t_1^*} f_0[x^*(t), u^*(t)] \, dt = \mathcal{V}(x^0). \tag{15.3}$$

Chap. 15 • Sufficient Conditions

Now consider

$$\Delta \triangleq \int_{t_0}^{t_1} f_0[x(t), u(t)]\, dt - \int_{t_0}^{t_1^*} f_0[x^*(t), u^*(t)]\, dt.$$

Then it follows from (15.3) that

$$\Delta = \int_{t_0}^{t_1} f_0[x(t), u(t)]\, dt - \mathcal{V}(x^0),$$

and finally from (15.2), together with the condition (iii) of the theorem, we arrive at

$$\Delta \geq \lim_{\bar{t} \to t_1} \int_{t_0}^{\bar{t}} \{ f_0[x(t), u(t)] + \operatorname{grad}^T \mathcal{V}[x(t)]\, f[x(t), u(t)] \}\, dt$$
$$\geq 0.$$

This concludes the proof. □

Before illustrating the use of Theorem 15.1 we make the following remarks. As we noted already earlier, the theorem embodies *local* sufficiency in that $u^*(\cdot)$ is compared only with controls which generate trajectories lying in X. Of course, if $X = R^n$ — or even if $X \cup \theta^1 = E$, the set of states from which the given target set θ^1 can be reached — then the comparison is with *all* controls which are feasible at x^0 and hence the sufficiency is *global*. An important consideration is the determination of a *test function* $\mathcal{V}(\cdot)$. In view of (15.3), the restriction to X of the optimal cost function, $V^*(\cdot)$, is an appropriate test function provided it is of class C^1; that is,

$$\mathcal{V}(\cdot) = V^*(\cdot)|_X.$$

Indeed Theorem 15.1 is called a *field theorem* because the determination of $V^*(\cdot)|_X$ requires the "field" of optimal trajectories which emanate from the points of X. Thus, if we wish to employ Theorem 15.1 in order to ascertain that a control $u(\cdot)$ which is extremal at x^0 is indeed optimal at x^0 with respect to X, we must obtain an extremal control for every $x^0 \in X$ so as to permit the construction of $\mathcal{V}(\cdot)$ by means of

$$\mathcal{V}(x^0) = \int_{t_0}^{t_1} f_0[x(t), u(t)]\, dt.$$

Finally, as mentioned in Section 15.1, the application of the sufficient conditions to the various generalizations discussed in Chapter 13 is easily accomplished. However, to do so one requires "optimality on θ^0" rather than just "optimality at x^0." Thus, if the initial state is not prescribed but merely restricted to belong to a given set θ^0, we modify the definition of "optimality at x^0 with respect to X." Given an initial state set $\theta^0 \subset R^n$ and a set X, $\theta^0 \cap X \neq \varnothing$, the control $u^*(\cdot):[t_0, t_1^*] \to R^m$, generating the solution $x^*(\cdot):[t_0, t_1^*] \to R^n$, is optimal on θ^0 with respect to X if and only if $u^*(\cdot) \in \mathcal{U}[x^*(t_0)]$, $x^*(t) \in X$ $\forall t \in [t_0, t_1^*)$ and

$$V[x^*(t_0), u^*(\cdot), x^*(\cdot)] \leq V[x^0, u(\cdot), x(\cdot)]$$

for every control $u(\cdot) \in \mathcal{U}(x^0)$ generating the solution $x(\cdot):[t_0, t_1] \to R^n$, $x(t_0) = x^0$, and $x(t) \in X$ $\forall t \in [t_0, t_1)$, and for all $x^0 \in \theta^0 \cap X$.

Now it is easily shown that the following condition, together with (i)–(iii) of Theorem 15.1, assures that $u^*(\cdot)$ is optimal on θ^0 with respect to X:

(iv) $\qquad\qquad \mathcal{V}[x^*(t_0)] \leq \mathcal{V}(x) \qquad \forall x \in \theta^0 \cap X.$

The proof is left to the reader (Exercise 15.1).

15.3. An Illustrative Example

Consider the state equations

$$\dot{x}_1(t) = 1 + u_1(t), \qquad \dot{x}_2(t) = u_2(t), \tag{15.4}$$

with the control subject to the constraint

$$u_1^2(t) + u_2^2(t) = 1. \tag{15.5}$$

It is desired to transfer the state from a given initial one, x^0, to $x^1 = 0$ while minimizing the value of

$$t_1 - t_0 = \int_{t_0}^{t_1} dt. \tag{15.6}$$

It is left as an exercise for the reader (Exercise 15.2) to show that the extremal control $u(\cdot):[t_0, t_1] \to R^2$ is given by

$$u_1(t) \equiv -1 - \frac{x_1^0}{\tau_1}, \qquad u_2(t) \equiv -\frac{x_2^0}{\tau_1} \tag{15.7}$$

where

$$\tau_1 \triangleq t_1 - t_0 = -\frac{(x_1^0)^2 + (x_2^0)^2}{2x_1^0}. \tag{15.8}$$

It is also readily verified that the set of initial states from which $x^1 = 0$ can be reached by means of an admissible control is

$$E = \{x \in R^2 | x_1 < 0\} \cup \{x^1\}. \tag{15.9}$$

In order to utilize Theorem 15.1 we require a test function $\mathcal{V}(\cdot)$. As noted in Section 15.2, we are led to the function defined by the cost due to an extremal control for every $x^0 \in X$. Here this is an easy task in view of (15.8); namely we have

$$\mathcal{V}(x) = -\frac{x_1^2 + x_2^2}{2x_1} \tag{15.10}$$

for all $x \in X = E \setminus \{x^1\}$. The function $\mathcal{V}(\cdot)$ satisfies the requirements of the theorem; that is, it is of class C^1 and meets the condition (i). The condition (ii) is verified by direct substitution of $u(t)$ given by (15.7) and by noting that

$$u_1(t) \equiv -1 - \frac{x_1^0}{\tau_1} = -1 - \frac{x_1(t)}{t_1 - t},$$

$$u_2(t) \equiv -\frac{x_2^0}{\tau_1} = -\frac{x_2(t)}{t_1 - t}.$$

Finally one can verify condition (iii) by showing that for $x \in X$ we have

$$\min_{u \in U} \left[1 + \frac{\partial \mathcal{V}(x)}{\partial x_1}(1 + u_1) + \frac{\partial \mathcal{V}(x)}{\partial x_2} u_2 \right] = 0.$$

Thus we can conclude that the extremal control given by (15.7) is indeed optimal at x^0, and furthermore that the minimum is global since $X = E \setminus \{x^1\}$.

15.4. Another Field Theorem

The sufficient conditions embodied in Theorem 15.1 involve a test function, $\mathcal{V}(\cdot): X \to R^1$, that is of class C^1. As we noted earlier, this test function is the restriction to X of the optimal cost function $V^*(\cdot)$. Consequently, in order to utilize the theorem, one must have an optimal cost function such that $V^*(\cdot)|_X$ is continuously differentiable. This requirement is unduly restrictive in many problems; for instance see the example of Section 15.5.

In this section we state a sufficiency theorem that involves a test function which is only "piecewise of class C^1." Before giving a definition of this class of functions we introduce the notion of a *finite decomposition* of a subset of R^n. A *finite decomposition* of a set $X \subset R^n$ is a finite collection of pairwise disjoint subsets X_i of X whose union is X. In other words, the finite decomposition D of X is defined as

$$D \triangleq \{X_i \subset X \mid i = 1, 2, \ldots, \kappa < \infty\}$$

such that

$$X_i \cap X_j = \emptyset \quad \text{if } i \neq j,$$

and

$$\bigcup_{i=1}^{\kappa} X_i = X.$$

Now we are ready to define the term "piecewise of class C^1." Let X be a subset of R^n and let D be a finite decomposition of X. Then a continuous function $\mathcal{V}(\cdot): X \to R^1$ is *piecewise of class C^1 with respect to D* if and only if there exists a collection

$$\{W_i, \mathcal{V}_i(\cdot) \mid i = 1, 2, \ldots, \kappa\}$$

such that W_i is an open set containing X_i and $\mathcal{V}_i(\cdot): W_i \to R^1$ is of class C^1, and $\mathcal{V}_i(x) = \mathcal{V}(x)$ for $x \in X_i$. Loosely speaking, a continuous function $\mathcal{V}(\cdot)$ is piecewise of class C^1 if it "agrees" on each subset of a finite decomposition of X with a function of class C^1.

Now we have the following theorem.

Theorem 15.2. *The control $u^*(\cdot): [t_0, t_1^*] \to R^m$ generating the solution $x^*(\cdot): [t_0, t_1^*] \to R^n$, $x^*(t_0) = x^0$, is optimal at x^0 with respect to $X \subset R^n$ if there exists a finite decomposition $D = \{X_i \mid i = 1, 2, \ldots, \kappa\}$ of X and a continuous function $\mathcal{V}(\cdot): X \to R^1$ that is piecewise of class C^1 with respect to D such that*

(i) *if $u(\cdot) \in \mathcal{U}(x^0)$ generates the solution $x(\cdot): [t_0, t_1] \to R^n$, $x(t_0) = x^0$, $x(t_1) \in \theta^1$ and $x(t) \in X$ for all $t \in [t_0, t_1)$, then*

$$\lim_{t \to t_1} \mathcal{V}[x(t)] \leq \lim_{t \to t_1^*} \mathcal{V}[x^*(t)] = 0;$$

and for $i \in \{1, 2, \ldots, \kappa\}$,
   (ii) $f_0[x^*(t), u^*(t)] + \text{grad}^T \tilde{V}_i[x^*(t)] f[x^*(t), u^*(t)] = 0$
   $\forall t \in \{t \in [t_0, t_1^*) | x^*(t) \in X_i\}$;
   (iii) $f_0(x, u) + \text{grad}^T \tilde{V}_i(x) f(x, u) \geq 0 \qquad \forall x \in X_i, \forall u \in U$.

Proof. The proof of this theorem is beyond the scope of this book. However Theorem 15.2 is a special case of the sufficiency theorem stated and proved in Ref. 15.1. □

We see that the conditions of Theorem 15.2 are analogous to those of Theorem 15.1, except that the conditions (ii) and (iii) apply separately for each member of the decomposition of X.

Finally we note that both Theorems 15.1 and 15.2 are readily modified to allow for state-dependent control constraints, as in the type of problems discussed in Section 13.10, by replacing U by $\hat{U}(x)$ and employing "admissible at x^0" in place of "admissible" controls (Exercise 15.4).

15.5. Time-Optimality for a Constant-Power Rocket

Let us recall the problem introduced in Section 12.1. We showed there that the initial state region is

$$E = \{x \in R^2 | x_2 < x_2^1\} \cup \{x^1\}, \tag{15.11}$$

and we deduced the extremal control $u(\cdot)$ given by

$$|u(t)| \equiv 1 \qquad \text{for } x^0 \in \text{I} \tag{15.12}$$

where

$$\text{I} \triangleq \{x \in R^2 | x_2 - x_2^1 \leq -|x_1 - x_1^1|\}, \tag{15.13}$$

and

$$u(t) \equiv \frac{x_2^1 - x_2^0}{x_1^1 - x_1^0} \qquad \text{for } x^0 \in \text{II} \cup \text{III} \tag{15.14}$$

where

$$\text{II} \triangleq \{x \in R^2 | x_1 < x_1^1, x_1 - x_1^1 < x_2 - x_2^1 < 0\},$$

$$\text{III} \triangleq \{x \in R^2 | x_1 > x_1^1, -(x_1 - x_1^1) < x_2 - x_2^1 < 0\}. \tag{15.15}$$

We also found the values of the cost corresponding to the extremal control; namely,

$$t_1 - t_0 = x_2^1 - x_2^0 \quad \text{for } x^0 \in \text{I},$$

$$t_1 - t_0 = \frac{(x_1^1 - x_1^0)^2}{x_2^1 - x_2^0} \quad \text{for } x^0 \in \text{II} \cup \text{III}. \tag{15.16}$$

Now consider the test function $\mathcal{V}(\cdot): X \to R^1$ chosen in conformity with (15.16), that is, such that

$$\mathcal{V}(x) = \begin{cases} x_1^1 - x_2 & \text{for } x \in \text{I} \cap X, \\ \dfrac{(x_1^1 - x_1)^2}{x_2^1 - x_2} & \text{for } x \in [\text{II} \cup \text{III}] \cap X. \end{cases} \tag{15.17}$$

Then it is readily shown (recall Exercise 12.1) that grad $\mathcal{V}(x)$ is not defined for $x \in \text{I} \cap \overline{\text{II}}$ and $x \in \text{I} \cap \overline{\text{III}}$, since

$$\text{grad } \mathcal{V}(x) = \begin{bmatrix} 0 \\ -1 \end{bmatrix} \quad \text{for } x \in \text{I} \cap X,$$

$$\text{grad } \mathcal{V}(x) = \begin{bmatrix} -2\dfrac{x_1^1 - x_1}{x_2^1 - x_2} \\ \left(\dfrac{x_1^1 - x_1}{x_2^1 - x_2}\right)^2 \end{bmatrix} \quad \text{for } x \in [\text{II} \cup \text{III}] \cap X.$$

Thus the function $\mathcal{V}(\cdot)$ is not of class C^1 on $X = E \setminus \{x^1\}$. Furthermore, as we noted in Section 12.1, the terminal portion of every trajectory in R^2 generated by the extremal control $u(\cdot)$ belongs to $\text{I} \cap \overline{\text{II}}$ or to $\text{I} \cap \overline{\text{III}}$ if $x^0 \in \text{I}$. Consequently Theorem 15.1 is not suitable. However, Theorem 15.2 can be utilized. To do so consider the decomposition $D = \{X_i | i = 1, 2, 3\}$ of $X = E \setminus \{x^1\}$ given by

$$X_1 = \text{I} \setminus \{x^1\}, \quad X_2 = \text{II}, \quad X_3 = \text{III}.$$

Also let the $\mathcal{V}_i(\cdot): W_i \to R^1$ be such that

$$\mathcal{V}_1(x) = x_2^1 - x_2, \quad W_1 = X,$$

$$\mathcal{V}_2(x) = \mathcal{V}_3(x) = \frac{(x_1^1 - x_1)^2}{x_2^1 - x_2}, \quad W_2 = X_2, \quad W_3 = X_3.$$

Now it is easily seen that $\mathcal{V}(\cdot)$, given by (15.17), is continuous on X and that it satisfies the condition (i). The condition (ii) of Theorem 15.2 is verified by the direct substitution of (15.12) and (15.14), respectively; namely,

$$|u(t)| = 1 \qquad \text{for } x(t) \in X_1,$$

$$u(t) = \frac{x_2^1 - x_2(t)}{x_1^1 - x_1(t)} \qquad \text{for } x(t) \in X_2 \cup X_3.$$

The latter relation follows from

$$\frac{x_2^1 - x_2(t)}{x_1^1 - x_1(t)} \equiv \frac{x_2^1 - x_2^0}{x_1^1 - x_1^0} \qquad \text{for } x^0 \in \text{II} \cup \text{III}.$$

The condition (iii) of Theorem 15.2 is verified by checking the sign of

$$1 + [u \quad u^2] \operatorname{grad} \mathcal{V}_i(x).$$

Thus, for $x \in X_1$, we have

$$1 + [u \quad u^2] \begin{bmatrix} 0 \\ -1 \end{bmatrix} = 1 - u^2 \geq 0,$$

while, for $x \in X_2 \cup X_3$, we have

$$1 + [u \quad u^2] \begin{bmatrix} -2\dfrac{x_1^1 - x_1}{x_2^1 - x_2} \\ \left(\dfrac{x_1^1 - x_1}{x_2^1 - x_2}\right)^2 \end{bmatrix} = \left[1 - u\dfrac{x_1^1 - x_1}{x_2^1 - x_2}\right]^2 \geq 0.$$

We conclude that the conditions of Theorem 15.2 are satisfied for $X = E \setminus \{x^1\}$ so that the extremal control is indeed (globally) optimal at x^0.

15.6. A Direct Sufficiency Theorem

In this section we derive sufficient conditions which differ in two respects from those given in Sections 15.2 and 15.4. The first difference arises from the kind of problems to which the sufficient conditions are applicable. Here we consider only problems in which the interval $[t_0, t_1]$ is precribed. The state equations may be nonautonomous; thus we allow problems of the type discussed in Sections 13.6 or 13.7 without resorting to equivalent systems. However, since $[t_0, t_1]$ is prescribed, the initial state and target sets do not depend on t; that is, they are given sets θ^0 and $\theta^1 \subset R^n$, respectively. The second way in which the *direct* sufficiency theorem differs from those of the *field* variety concerns the *test function*. Here the test function is from $[t_0, t_1]$ into R^n; that is, it is a function of the independent variable rather than of the state.

Before stating a direct sufficiency theorem we define the function $\mathcal{H}(\cdot): R^n \times R^n \times R^1 \times R^m \to R^1$ by

$$\mathcal{H}(p, x, t, u) \triangleq -\hat{f}_0(x, t, u) + p^T \hat{f}(x, t, u). \qquad (15.18)$$

Then we have the following theorem.

Theorem 15.3. *The control* $u^*(\cdot): [t_0, t_1] \to R^m$, *generating the solution* $x^*(\cdot): [t_0, t_1] \to R^n$, *is optimal on* θ^0 *with respect to* $X \subset R^n$ *if there exists a piecewise smooth function* $p(\cdot): [t_0, t_1] \to R^n$ *such that*

(i) $\mathcal{H}[p(t), x^*(t), t, u^*(t)] - \mathcal{H}[p(t), x, t, u] + \dot{p}^T(t)[x^*(t) - x] \geq 0$
$\forall x \in X, \forall u \in U$ *and* $\forall t \in [t_0, t_1]$;

(ii) $p^T(t_0)[x^*(t_0) - y] - p^T(t_1)[x^*(t_1) - z] \geq 0$
$\forall y \in \theta^0 \cap X$ *and* $\forall z \in \theta^1 \cap \bar{X}$.

Proof. Let $u(\cdot): [t_0, t_1] \to R^m$, generating the solution $x(\cdot): [t_0, t_1] \to R^n$, $x(t_0) \in \theta^0 \cap X$, be feasible at $x(t_0)$ and such that $x(t) \in X$ for all $t \in [t_0, t_1)$. As a consequence of the condition (i) of the theorem and of the state equations (13.35) we have

$$\hat{f}_0[x(t), t, u(t)] - \hat{f}_0[x^*(t), t, u^*(t)] - \frac{d}{dt}\{p^T(t)[x(t) - x^*(t)]\} \geq 0.$$

Hence, after integration, the condition (ii) implies that

$$\int_{t_0}^{t_1} \hat{f}_0[x(t), t, u(t)] \, dt - \int_{t_0}^{t_1} \hat{f}_0[x^*(t), t, u^*(t)] \, dt \geq 0.$$

This concludes the proof. □

Unlike in the field theorems in which the test function $\mathcal{V}(\cdot)$ is obtained from the optimal cost function $V^*(\cdot)$, here we have not given a recipe for determining a test function $p(\cdot)$. However, if $x^*(t)$ belongs to the interior of X for all $t \in [t_0, t_1)$, then it follows from condition (i) of the theorem that $p(\cdot)$ is a solution of the adjoint equations

$$\dot{p}_i(t) = -\frac{\partial \mathcal{H}[p(t), x^*(t), t, u^*(t)]}{\partial x_i}, \quad i = 1, 2, \ldots, n.$$

Some extensions of Theorem 15.3, including the extension to state-dependent control constraints, can be found in Ref. 15.2. Other direct sufficiency theorems as well as a general discussion of this topic are in Ref. 15.3.

15.7. An Illustrative Example

Consider the state equations

$$\dot{x}_1(t) = u_1(t), \quad \dot{x}_2(t) = u_2(t). \tag{15.19}$$

The control is unconstrained; that is, $U = R^2$. The initial state and target sets are

$$\theta^0 = \{x \in R^2 \mid x_2^2 + x_1 = 0\}, \quad \theta^1 = \{x \in R^2 \mid x_2^2 - x_1 + 1 = 0\}.$$

It is desired to transfer the state from one in θ^0 to one in θ^1 while minimizing the value of the cost

$$\int_0^2 [u_1^2(t) + u_2^2(t)]^{1/2} \, dt.$$

Note that the interval is prescribed; namely, $[t_0, t_1] = [0, 2]$.

We leave it as an exercise for the reader to deduce the *extremal* control (Exercise 15.11); it is given by

$$u_1(t) \equiv \tfrac{1}{2}, \quad u_2(t) \equiv 0 \tag{15.20}$$

so that

$$x_1(t) = \tfrac{1}{2}t, \quad x_2(t) \equiv 0. \tag{15.21}$$

The corresponding solution of the adjoint equations is given by

$$\lambda_0(t) \equiv -1, \quad \lambda_1(t) \equiv 1, \quad \lambda_2(t) \equiv 0.$$

Now it is easy to verify the conditions of Theorem 15.3 for $X = R^2$ using the test function $p(\cdot)$ with

$$p_1(t) = \lambda_1(t) \equiv 1. \quad p_2(t) = \lambda_2(t) \equiv 0.$$

Thus, in view of the state equations (15.19) with the control (15.20), the condition (i) of Theorem 15.3 becomes

$$\left(u_1^2 + u_2^2\right)^{1/2} - u_1 \geq 0 \quad \forall u \in R^2.$$

As a result of the solution (15.21) and of the end conditions, whence

$$y_1 \leq 0, \quad z_1 \geq 1,$$

the condition (ii) is

$$(0 - y_1) - (1 - z_1) \geq 0.$$

We conclude that the extremal control is (globally) optimal on θ^0 with respect to $X = R^2$.

15.8. Life History Strategies of Plants

A commonly held view of the emergence of certain species of living organisms is that of the "survival of the fittest." In other words, natural selection operates through the evolution of life history strategies which maximize "fitness." That is not to say that an individual plant or animal necessarily adopts a policy by a *conscious* effort to maximize the value of some functional. Rather it means that an "optimal" policy *evolves* through a process of trial and error. Indeed it is the success of a particular policy that brands it as being "optimal." For an enlightening discussion of this topic see Ref. 15.4.

To illustrate the use of optimal control theory in the modeling of a biological system let us consider Cohen's problem dealing with asexual annual plants; see Ref. 15.5. The problem of interest is that of partitioning the photosynthetic products between vegetative and reproductive growth so as to maximize the seed yield. Here we consider a version of this problem as treated in Ref. 15.6. Let t be the time, $x_1(t)$ be the leaf biomass, $x_2(t)$ be the reproductive biomass (flowers, seeds), $x_3(t)$ be the remaining biomass (vegetative), $g_i(t)$ be the fraction of total growth due to photosynthesis, allocated to component $i = 1, 2, 3$ at time t, and r be the photosynthetic production per unit mass $= \text{const} > 0$.

Thus the rates of change of the biomasses are

$$\begin{aligned} \dot{x}_1(t) &= rg_1(t)x_1(t), \\ \dot{x}_2(t) &= rg_2(t)x_1(t), \\ \dot{x}_3(t) &= rg_3(t)x_1(t). \end{aligned} \quad (15.22)$$

Chap. 15 • Sufficient Conditions

Since the $g_i(t)$ are fractions of the total growth we have

$$g_i(t) \geq 0, \quad i = 1, 2, 3,$$
$$g_1(t) + g_2(t) + g_3(t) = 1. \tag{15.23}$$

In this model it is assumed that the leaf and vegetative biomasses are related by

$$\frac{x_1(t)}{x_1(t) + x_3(t)} \equiv \text{const} \triangleq l > 0. \tag{15.24}$$

On combining (15.22)–(15.24) one obtains

$$\dot{x}_1(t) = lr\big[1 - g_2(t)\big] x_1(t). \tag{15.25}$$

Assuming that "fitness" is proportional to the reproductive biomass at the end of a growing season of *fixed duration* $t_1 - t_0$, the quantity to be maximized is $x_2(t_1)$. By (15.22) we have

$$x_2(t_1) - x_2(t_0) = r \int_{t_0}^{t_1} g_2(t) x_1(t) \, dt.$$

Thus, for a given initial reproductive biomass $x_2(t_0)$, the quantity to be *minimized* is

$$-\int_{t_0}^{t_1} r g_2(t) x_1(t) \, dt. \tag{15.26}$$

Now we note the following. Only the *initial* values of the $x_i(t)$, $i = 1, 2, 3$, are specified. Furthermore the cost (15.26) as well as the equation for $x_1(\cdot)$, equation (15.25), involve only $x_1(t)$ and $g_2(t)$. Hence we are concerned with a single state variable, x_1, and a single control variable, g_2. Letting $g_2 = u$, we rewrite (15.25) as our only state equation

$$\dot{x}_1(t) = lr\big[1 - u(t)\big] x_1(t). \tag{15.27}$$

The initial state is prescribed, $x_1(t_0) = x_1^0 > 0$, as is the interval $t_1 - t_0$. The cost is

$$-\int_{t_0}^{t_1} r u(t) x_1(t) \, dt. \tag{15.28}$$

In view of (15.23) the control is constrained by

$$u(t) \in [0,1]. \tag{15.29}$$

Now we are ready to utilize the maximum principle, as modified for a fixed interval in Section 13.7, to deduce the extremal control. The function $H(\cdot)$ is given by

$$H(\lambda, y, u) = -\lambda_0 r u x_1 + \lambda_1 lr(1-u) x_1 \tag{15.30}$$

so that the adjoint equations are

$$\dot{\lambda}_0(t) = 0, \quad \dot{\lambda}_1(t) = \lambda_0(t) r u(t) - \lambda_1(t) lr[1-u(t)]. \tag{15.31}$$

Since $x_1(t_1)$ is not specified it follows from the terminal transversality condition that

$$\lambda_1(t_1) = 0. \tag{15.32}$$

The condition (c) of Theorem 11.1 and the requirement for a nonzero solution of (15.31) imply that $\lambda_0(t) \equiv \text{const} < 0$. We may choose

$$\lambda_0(t) \equiv -1. \tag{15.33}$$

Now it follows from (15.30) with (15.33) that

$$H[\lambda(t), y(t), u] = r x_1(t)[1 - l\lambda_1(t)] u + lr x_1(t) \lambda_1(t). \tag{15.34}$$

Of course one must have a nonnegative biomass. Since $x_1^0 > 0$ and $u(t) \in [0,1]$, the state equation (15.27) leads to

$$x_1(t) > 0 \quad \forall t \in [t_0, t_1].$$

Consequently the condition (a) of Theorem 11.1, applied to (15.34), results in the extremal control $u(\cdot): [t_0, t_1] \to R^1$ such that

$$u(t) = \begin{cases} 1 & \text{if } l\lambda_1(t) < 1, \\ 0 & \text{if } l\lambda_1(t) > 1. \end{cases} \tag{15.35}$$

It is readily shown that

$$l\lambda_1(t) \equiv 1 \tag{15.36}$$

Chap. 15 • Sufficient Conditions

on a nonzero subinterval of $[t_0, t_1]$ is not possible. For (15.36) implies that

$$\dot{\lambda}_1(t) = 0$$

while (15.31), together with (15.33) and (15.36), implies that

$$\dot{\lambda}_1(t) = -r.$$

In view of (15.32) and (15.35) we have $u(t_1) = 1$. Furthermore, by the continuity of $\lambda_1(\cdot)$, there is a $\Delta t > 0$ such that

$$l\lambda_1(t) < 1 \quad \forall t \in [t_1 - \Delta t, t_1]$$

so that

$$u(t) = 1 \quad \forall t \in [t_1 - \Delta t, t_1]. \tag{15.37}$$

On integrating (15.31) with (15.33) and (15.37) we obtain

$$\lambda_1(t) = r(t_1 - t) \quad \text{for } t \in [t_1 - \Delta t, t_1]. \tag{15.38}$$

The instant, t_s, when the control *switches* is readily found since

$$l\lambda_1(t_s) = 1,$$

so that (15.38) yields

$$1 = lr(t_1 - t_s),$$

that is,

$$t_1 - t_s = \frac{1}{lr}. \tag{15.39}$$

Provided $t_1 > t_0 + 1/lr$, one can be certain that switching takes place since

$$\dot{\lambda}_1(t_s \pm 0) = -r < 0$$

so that $1 - l\lambda_1(t)$ changes sign at $t = t_s$.

Continuing the backwards integration, now with $u(t) = 0$, yields

$$\lambda_1(t) = \lambda_1(t_s) e^{lr(t_s - t)} \quad \text{for } t < t_s. \tag{15.40}$$

Hence $\lambda_1(t)$ increases with decreasing t and no further sign change of $1 - l\lambda_1(t)$, and hence no further switch of the extremal control, is possible.

Thus the unique extremal control is such that

$$u(t)=0 \quad \text{for } t\in[t_0,t_s),$$
$$u(t)=1 \quad \text{for } t\in[t_s,t_1] \tag{15.41}$$

where, by (15.39), the switching time is

$$t_s = t_1 - \frac{1}{lr}. \tag{15.42}$$

Now we need to check whether the extremal control is indeed optimal at x_1^0. We can employ Theorem 15.3 since the interval $t_1 - t_0$ is fixed. Since our interest is in nonnegative biomass and $x_1^0 \geq 0$ implies that $x_1(t) \geq 0$, we take

$$X = \{x_1 \in R^1 | x_1 \geq 0\}.$$

As the test function $p(\cdot)$ we choose $p(\cdot) = \lambda_1(\cdot)$. To verify the condition (i) of Theorem 15.3 we must check the sign of

$$rx_1(t)u(t) + \lambda_1(t)lrx_1(t)[1-u(t)] - rx_1 u - \lambda_1(t)lrx_1(1-u)$$
$$- \{ru(t) + \lambda_1(t)lr[1-u(t)]\}[x_1(t) - x_1],$$

which reduces to

$$rx_1[1 - l\lambda_1(t)][u(t) - u]. \tag{15.43}$$

As a consequence of (15.35) and the constraint (15.29), the expression (15.43) is nonnegative so that the condition (i) is verified. The condition (ii) is satisfied since $\theta^0 = \{x_1^0\}$ and $\lambda_1(t_1) = 0$. Consequently we conclude that the extremal control (15.41) is globally optimal for $x_1^0 \geq 0$.

In other words, based on Cohen's model of an asexual annual plant, we conclude that a plant puts no effort into seed production until $1/lr$ units of time before the end of the growing season. Such a policy is "optimal" from the point of view of maximum "fitness."

15.9. An Economic Control Problem

Now we turn to a discussion of the optimal maintenance policy for rental housing. In particular we consider a version of the problem treated in Ref. 15.7. A landlord wishes to maximize his long-term, possibly discounted

Chap. 15 • Sufficient Conditions

net profit through the proper management of his property by means of the maintenance policy which affects both his short-term expenses and his long-term profit, the former because of associated costs and the latter because of increased rent for better housing.

Let t be the time, x_1 be the quality level of housing, α be the quality of new or top condition housing = const > 0, and u be the maintenance control.

The state equation is taken as

$$\dot{x}_1(t) = -\beta(t)x_1(t) + u(t) - \frac{1}{2}\frac{c(t)u^2(t)}{\alpha - x_1(t)} \qquad (15.44)$$

over the *fixed interval* $[t_0, t_1]$, where $\beta(\cdot)$ and $c(\cdot)$ from $[t_0, t_1] \to R^1$ are functions of class C^1 with bounded positive values.

The term $-\beta(t)x_1(t)$ represents the effect of decline in the absence of maintenance, while the term $u(t) - \frac{1}{2}[c(t)u^2(t)][\alpha - x_1(t)]^{-1}$ represents the effect of maintenance. If $u(t) \equiv 0$, the quality level declines; for instance, if $\beta(t) \equiv$ constant, the decline is exponential. The maintenance effort increases the quality level but with a diminishing return; the effect decreases as the top condition is approached. Once the maintenance control $u(\cdot)$ reaches the value

$$u(t) = \frac{\alpha - x_1(t)}{c(t)},$$

corresponding to the maximum of the maintenance effort, no further increase seems desirable since a larger value of $u(t)$ would tend to decrease $\dot{x}_1(t)$.

Of course both $x_1(t)$ and $u(t)$ must be nonnegative. Thus there are the constraints

$$0 \leq x_1(t) \leq \alpha \qquad (15.45)$$

and

$$0 \leq u(t) \leq \frac{\alpha - x_1(t)}{c(t)}. \qquad (15.46)$$

The former is a constraint on the state, while the latter is a state-dependent control constraint. In view of (15.46), $u^2(t) \to 0$ more rapidly than does $[\alpha - x_1(t)]/c(t)$ so that the maintenance effort goes to zero as $x_1(t) \to \alpha$. Also we shall see shortly that the state constraint (15.45) is always satisfied provided it is met at $t=t_0$. Furthermore the state-dependent upper bound on $u(t)$ need not be imposed; as will be shown, the optimal control $u^*(\cdot)$, in the *larger* class of admissible controls *without* that upper bound, does not exceed that bound.

We prove now that (15.45) is always satisfied. Suppose that $x_1(t_0) = x_1^0 \geq 0$ and that there is a $\tau \in [t_0, t_1]$ such that $x_1(\tau) = 0$. Then

$$\dot{x}_1(\tau) = u(\tau) - \frac{1}{2}\frac{c(\tau)u^2(\tau)}{\alpha} \geq 0$$

for all $u(\tau)$ satisfying (15.46). Thus $x_1(\cdot)$ cannot decrease further; that is,

$$x_1(t) \geq 0 \qquad \forall t \in [t_0, t_1].$$

Now suppose that $x_1(t_0) = x_1^0 \leq \alpha$ and that there is a $\tau \in [t_0, t_1]$ such that $x_1(\tau) = \alpha$. Then

$$\dot{x}_1(\tau) = -\beta(\tau)\alpha < 0$$

since

$$u(t) - \frac{1}{2}\frac{c(t)u^2(t)}{\alpha - x_1(t)} \to 0 \qquad \text{as } x_1(t) \to \alpha.$$

Thus $x_1(t)$ cannot exceed $x_1(\tau) = \alpha$ for $t > \tau$, and hence

$$x_1(t) \leq \alpha \qquad \forall t \in [t_0, t_1].$$

Consequently we need not impose the state constraint (15.45) provided, as we suppose,

$$0 \leq x_1^0 \leq \alpha.$$

We impose but a single constraint, namely, the control constraint

$$0 \leq u(t). \tag{15.47}$$

The *net* profit over the given period is

$$\int_{t_0}^{t_1} [\gamma(t)x_1(t) - s(t)u(t)]\, dt \tag{15.48}$$

where $\gamma(\cdot)$ and $s(\cdot)$ from $[t_0, t_1] \to R^1$ are given functions of class C^1 with positive values. The term $\gamma(t)x_1(t)$ is the rate of rental income at time t; it depends on the quality level of the housing. The term $s(t)u(t)$ is the rate of maintenance expenditures; it depends on the maintenance effort. Here one can account for discounting through the time-dependence of $\gamma(\cdot)$ and $s(\cdot)$; that is, future income as well as expenditures are worth less than current ones. Since one wishes to *maximize* the net profit, the cost to be *minimized* is the negative value of (15.48). Note that we have an optimal control problem for a nonautonomous system since the independent variable, here the time t, appears explicitly in the state equation and in the cost integrand as well as in the end conditions because the interval is prescribed. Hence we must invoke the necessary conditions as modified in Section 13.6.

The function $\hat{H}(\cdot)$ is given by

$$\hat{H}(\lambda, y, t, u) = -\lambda_0[\gamma(t)x_1 - s(t)u] + \lambda_1\left[-\beta(t)x_1 + u - \frac{c(t)u^2}{2(\alpha - x_1)}\right]$$

(15.49)

so that the adjoint equations become

$$\dot{\lambda}_0(t) = 0, \quad \dot{\lambda}_1(t) = \lambda_0(t)\gamma(t) + \lambda_1(t)\beta(t) + \frac{\lambda_1(t)c(t)u^2(t)}{2[\alpha - x_1(t)]^2}.$$

(15.50)

Since $x_1(t_1)$ is not prescribed, the terminal transversality condition yields

$$\lambda_1(t_1) = 0. \tag{15.51}$$

Thus we conclude at once that $\lambda_0(t) \equiv \text{constant} \neq 0$, lest the solution of (15.50) is the trivial one. We may take $\lambda_0(t) \equiv -1$ as a result of the condition (c) of Theorem 11.1.

Next we show that

$$\lambda_1(t) > 0 \quad \forall t \in [t_0, t_1). \tag{15.52}$$

For it follows from (15.50) that if there is a $\tau \in [t_0, t_1)$ such that

$$\lambda_1(\tau) \leq 0,$$

then

$$\dot{\lambda}_1(\tau) < 0.$$

But that cannot be since then

$$\lambda_1(t) < 0 \quad \forall t \geq \tau,$$

which contradicts (15.51).

Next we invoke the condition (a) of Theorem 11.1. The part of $\hat{H}(\cdot)$ that depends on u is

$$[\lambda_1(t) - s(t)]u - \frac{\lambda_1(t) c(t)}{2[\alpha - x_1(t)]} u^2.$$

Consequently we conclude that an extremal control $u(\cdot)$ is such that

$$u(t) = \begin{cases} 0 & \text{if } 0 \leq \lambda_1(t) \leq s(t), \\ \left[1 - \frac{s(t)}{\lambda_1(t)}\right] \frac{\alpha - x_1(t)}{c(t)} & \text{if } s(t) < \lambda_1(t). \end{cases} \quad (15.53)$$

In other words, the value of the extremal control depends only on $\lambda_1(t)$. Now we note that the adjoint equations are decoupled from the state equation. For if

$$u(t) = 0$$

then

$$\dot{\lambda}_1(t) = -\gamma(t) + \lambda_1(t)\beta(t), \quad (15.54)$$

and if

$$u(t) = \left[1 - \frac{s(t)}{\lambda_1(t)}\right] \frac{\alpha - x_1(t)}{c(t)}$$

then

$$\dot{\lambda}_1(t) = -\gamma(t) + \lambda_1(t) \left\{ \beta(t) + \frac{1}{2c(t)} \left[1 - \frac{s(t)}{\lambda_1(t)}\right]^2 \right\}. \quad (15.55)$$

Thus one can integrate (15.54) backwards over the interval $[t_s, t_1]$ with $\lambda_1(t_1) = 0$ and t_s such that

$$\lambda_1(t_s) = s(t_s). \tag{15.56}$$

This is followed by the integration of (15.55) over the interval $[t_0, t_s]$ with $\lambda_1(t_s)$ given by (15.56). Of course if $t_s \leq t_0$, the integration of (15.55) ends at $t = t_0$ so that $u(t) \equiv 0$.

Since the interval $[t_0, t_1]$ is prescribed we may employ Theorem 15.3 in order to establish the optimality of the extremal control at $x_1^0 \in [0, \alpha]$ with respect to

$$X = \{x_1 \in R^1 | 0 \leq x_1 \leq \alpha\}.$$

We choose the test function $p(\cdot) = \lambda_1(\cdot)$. To verify the condition (i) of Theorem 15.3 we must consider the two intervals, $[t_0, t_s]$ and $[t_s, t_1]$, corresponding to $u(t)$ and $\lambda_1(t)$ given by (15.53). For instance, on the interval $[t_s, t_1]$ we have

$$u(t) = 0, \qquad \lambda_1(t) \in [0, s(t)].$$

Thus we must check the sign of

$$\gamma(t) x_1(t) - \lambda_1(t) \beta(t) x_1(t) - \gamma(t) x_1 + s(t) u$$

$$-\lambda_1(t) \left[-\beta(t) x_1 + u - \frac{c(t) u^2}{2(\alpha - x_1)} \right] + [-\gamma(t) + \lambda_1(t) \beta(t)][x_1(t) - x_1].$$

The value of this expression is nonnegative and hence the condition (i) is verified. We leave it as an exercise for the reader (Exercise 15.12) to verify the condition (i) on the interval $[t_0, t_s]$ where

$$u(t) = \left[1 - \frac{s(t)}{\lambda_1(t)} \right] \frac{\alpha - x_1(t)}{c(t)}, \qquad \lambda_1(t) > s(t).$$

The condition (ii) is satisfied since $\theta^0 = \{x_1^0\}$ and $\lambda_1(t_1) = 0$. We conclude that the extremal control $u(\cdot)$ is indeed optimal at x_1^0 with respect to X.

Finally we show that the control $u(\cdot)$ given by (15.53) satisfies the constraint

$$u(t) \le \frac{\alpha - x_1(t)}{c(t)}.$$

This follows at once from

$$s(t) < \lambda_1(t).$$

Thus the control $u(\cdot)$ is also optimal at x_1^0 with respect to X in the class of admissible controls constrained by (15.46). Furthermore, for such controls, all trajectories remain in X so that the control (15.53) is globally optimal at x_1^0.

Exercises

15.1. Show that the condition (iv) in Section 15.2, when added to the conditions (i)–(iii), modifies Theorem 15.1 to allow for optimality on θ^0.

15.2. Recall the example of Section 15.3. Deduce the extremal control.

15.3. Restate Theorems 15.1 and 15.2 for a nonautonomous system by employing an equivalent system as in Section 13.6.

15.4. Show how Theorems 15.1–15.3 are altered if the control constraint set is state-dependent.

15.5. Recall the minimum distance problem of Section 13.3. Use a sufficiency theorem to show that the straight line joining the prescribed curves orthogonally is indeed the curve of minimum length.

15.6. Recall the problem of Section 13.8. Use Theorem 15.1, as modified in Exercise 15.3, to show that the extremal control given by (13.67) is optimal at $\tilde{x}^0 = [x_1^0 \, x_2^0 \, t_0]^T$ with respect to $X \subset R^3$. What is X here?

15.7. Employ a field theorem, modified to allow for a state-dependent control constraint set, to prove that the extremal control in Section 13.11 is indeed optimal at x^0 with respect to X. What is X here?

15.8. Recall Dido's problem in Section 13.13. Use Theorem 15.1 to prove that Dido's solution renders the maximum enclosed area.

15.9. Recall the problem of Section 13.15. Pose this problem for an equivalent system, as in Section 13.14, and employ Theorem 15.3 to prove the optimality of the extremal control-parameter pair.

Chap. 15 • Sufficient Conditions

15.10. Recall the time-optimal regulator problem of Section 14.6. Utilize Theorem 15.2 to show that the extremal control is indeed optimal at x^0 with respect to $X = R^2 \setminus \{x^1\}$.

15.11. Recall the example of Section 15.7. Deduce the extremal control.

15.12. Recall the example of Section 15.9. Show that the condition (i) of Theorem 15.3 is satisfied for $t \in [t_0, t_s]$.

15.13. Recall the time-optimal navigation problem of Section 12.2. Show that the extremal control is optimal at x^0.

15.14. Show that

$$\underset{u \in U}{\text{Max}}\, \mathcal{H}[p(t), x^*(t), t, u] = \mathcal{H}[p(t), x^*(t), t, u^*(t)]$$

is a necessary condition for (i) of Theorem 15.3.

16

Feedback Control

16.1. Introduction

Let us recall the problem of optimal control defined in Chapter 10. The target set θ^1 is prescribed once and for all; it is one of the givens of the problem. A control that is optimal at the initial state x^0 must transfer the state to one in θ^1 as well as render the minimum value of the cost. We speak of a control $u^*(\cdot):[t_0, t_1^*] \to R^m$ as being *optimal at* x^0 in order to emphasize that the function $u^*(\cdot)$ is optimal if the initial state is x^0. In general, for a different initial state a different control function is optimal.

One reason for being concerned with the dependence of the optimality of a control on the initial state is the fact that real systems are not perfect; the actual state of the system may differ from that predicted by the solution of the state equations. To illustrate this consider the time-optimal regulator problem discussed in Section 14.6. Suppose that the initial state lies in the region below the switching curve AOB so that the control $u^*(\cdot):[t_0, t_1^*] \to R^1$, which is optimal at x^0, is such that

$$\begin{aligned} u^*(t) &= 1 & \text{for } t \in [t_0, t_s), \\ u^*(t) &= -1 & \text{for } t \in [t_s, t_1^*]. \end{aligned} \quad (16.1)$$

Let $x^*(\cdot):[t_0, t_1^*] \to R^2$, $x^*(t_0) = x^0$, denote the solution generated by $u^*(\cdot)$. Then it is readily shown (Exercise 14.1) that the instant of switching, t_s, is given by

$$t_s - t_0 = x_2^*(t_s) - x_2^0$$

where

$$x_2^*(t_s) = \left[\tfrac{1}{2}(x_2^0)^2 - x_1^0\right]^{1/2}.$$

Suppose that the control $u^*(\cdot)$ is employed but that the real system suffers a perturbation during the interval $[t_0, \bar{t}]$, $\bar{t} < t_s$, so that the actual state at time $\bar{t}$ is $\bar{x} \neq x^*(\bar{t})$. Suppose that no further perturbations occur, and let $x(\cdot): [\bar{t}, t_1^*] \to R^2$, $x(\bar{t}) = \bar{x}$, denote the solution generated by $u^*(\cdot)|_{[\bar{t}, t_1^*]}$. This is illustrated in Figure 16.1. Since the control $u^*(\cdot)$ is employed, the switch in its value from 1 to -1 occurs at time t_s. However, if one were to employ the control which is optimal at $\bar{x}$, the switch would occur at $\bar{t}_s$ given by

$$\bar{t}_s - \bar{t} = x_2(\bar{t}_s) - \bar{x}_2,$$

where

$$x_2(\bar{t}_s) = \left(\tfrac{1}{2}\bar{x}_2^2 - \bar{x}_1\right)^{1/2}.$$

Thus, in general, $t_s \neq \bar{t}_s$ and consequently $u^*(\cdot)|_{[\bar{t}, t_1^*]}$ is not optimal at $\bar{x}$. However, if one had implemented the switch at the instant the state reached

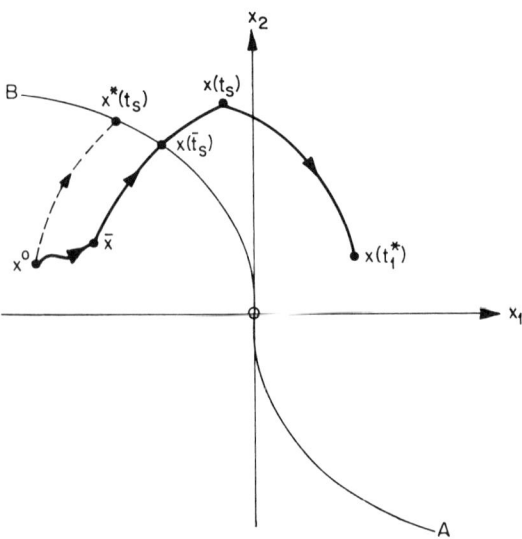

Figure 16.1. A perturbed system trajectory.

AOB, that is, based on the knowledge of the state rather than of the time—the resultant control would have been optimal at $\bar{x}$.

Because of considerations such as those discussed above, one is often interested in determining a function of the state, x, that permits one to determine the control which is optimal at that state. We are led to the following definition.

The function $k^*(\cdot): E^* \to R^m$ is an *optimal feedback control* if and only if, given $x^0 \in E^*$, there is a unique solution $x^*(\cdot): [t_0, t_1^*] \to R^n$ of

$$\dot{x}^*(t) = f[x^*(t), k^*(x^*(t))] \tag{16.2}$$

such that $x^*(t_0) = x^0$, $x^*(t_1^*) \in \theta^1$, and the function $u^*(\cdot): [t_0, t_1^*] \to R^m$ defined by

$$u^*(t) \triangleq k^*[x^*(t)], \quad t \in [t_0, t_1^*], \tag{16.3}$$

is optimal at x^0.

The appellation *feedback* is used because the state is "fed back" into the system equations. Such a control is also termed a *closed-loop* control because the loop between the output, here the state, and the input, here the control, is closed; see Figure 16.2. By contrast, the control $u^*(\cdot): [t_0, t_1^*] \to R^m$ is called an *open-loop* control.

There arise some questions concerning optimal feedback control. The first pertains to the existence and uniqueness of solutions of equation (16.2). Some aspects of this question will be considered briefly in Section 16.4. A second question concerns the existence, and perhaps more importantly the construction, of an optimal feedback control; some aspects of that question

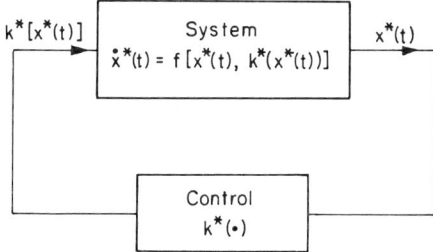

Figure 16.2. Feedback control.

are treated in Ref. 16.1. Let us turn first to the second question, that of the so-called *synthesis* of an optimal feedback control.

16.2. The Synthesis of Optimal Feedback Control

If, given $x^0 \in E^*$, there is a *unique* control $u^*(\cdot): [t_0, t_1^*] \to R^m$ that is optimal at x^0, then there exists an optimal feedback control $k^*(\cdot): E^* \to R^m$; it is defined by

$$k^*(x^0) = u^*(t_0), \qquad x^0 \in E^*. \tag{16.4}$$

Indeed this is the case in many problems treated earlier. In the problems of Sections 12.2, 14.6, and 14.10 the knowledge of the state determines the value of the open-loop control that is optimal at that state. For example, in the time-optimal regulator problem of Section 14.6, if

$$x^0 \in \{ x \in R^2 \mid x_1 > -\tfrac{1}{2} x_2 |x_2| \}$$

then $k^*(x^0) = 1$. In other problems, such as the one discussed in Section 13.8, the dependence of the optimal open-loop control, $u^*(\cdot)$, on the initial state, x^0, is *explicit*. Recall the optimal control for the problem of Section 13.8; it is given by

$$u^*(t) = \frac{2}{(t_1 - t_0)^2} \left\{ [3x_1^0 + x_2^0(t_1 - t_0)] - [6x_1^0 + 3x_2^0(t_1 - t_0)] \frac{t_1 - t}{t_1 - t_0} \right\}.$$

Since the system is nonautonomous we consider the equivalent system with the augmented state $\tilde{x} \triangleq [x_1 \ x_2 \ x_3]^T$, where $x_3 = t$. Thus we have

$$k^*(\tilde{x}^0) = \frac{-2}{(t_1 - t_0)^2} [3x_1^0 + 2x_2^0(t_1 - t_0)]$$

for all

$$\tilde{x}^0 \triangleq [x_1^0 \ x_2^0 \ t_0]^T \in \{ \tilde{x} \in R^3 \mid t < t_1 \}.$$

Another way to synthesize an optimal feedback control $k^*(\cdot): E^* \to R^m$ is to employ sufficient conditions of the field variety such as Theorems 15.1 and 15.2. For instance, consider Theorem 15.1 with $X = E^*$, and suppose there exists a function

$$\mathcal{V}(\cdot): E^* \to R^1$$

of class C^1 satisfying

$$\lim_{\substack{x \to x^1 \\ x \in E^*}} \mathcal{V}(x) = 0 \quad \forall x^1 \in \theta^1, \tag{16.5}$$

and a continuous function

$$\omega(\cdot): E^* \times R^n \to R^m,$$

such that for all $x \in E^*$ and $\psi \in R^n$

$$\min_{u \in U} \left[f_0(x, u) + \psi^T f(x, u) \right] = f_0[x, \omega(x, \psi)] + \psi^T f[x, \omega(x, \psi)], \tag{16.6}$$

and for all $x \in E^*$

$$f_0[x, \omega(x, \operatorname{grad} \mathcal{V}(x))] + \operatorname{grad}^T \mathcal{V}(x) f[x, \omega(x, \operatorname{grad} \mathcal{V}(x))] = 0. \tag{16.7}$$

The equation (16.7), often referred to as the Hamilton–Jacobi–Bellman equation, is a partial differential equation whose solution $\mathcal{V}(\cdot)$ is subject to the boundary condition (16.5). Now consider the function $k^*(\cdot): E^* \to R^m$ such that

$$k^*(x) = \omega[x, \operatorname{grad} \mathcal{V}(x)], \quad x \in E^*, \tag{16.8}$$

and suppose that, given $x^0 \in E^*$, there is a unique solution $x^*(\cdot): [t_0, t_1^*] \to R^n$ of equation (16.2) such that $x^*(t_0) = x^0$ and $x^*(t_1^*) \in \theta^1$. Then the control $u^*(\cdot): [t_0, t_1^*] \to R^m$, defined by (16.3), is optimal at x^0, since the conditions of Theorem 15.1 are satisfied. Thus the feedback control, defined by (16.8), is optimal.

16.3. The Linear-Quadratic Problem

Consider the state equation

$$\dot{x}(t) = A(t)x(t) + B(t)u(t) \tag{16.9}$$

with the initial state $x(t_0) = x^0$ and the interval $[t_0, t_1]$ specified. The cost to be minimized is

$$\frac{1}{2}\int_{t_0}^{t_1}\left[x^T(t)Q(t)x(t)+u^T(t)R(t)u(t)\right]dt. \qquad (16.10)$$

The elements of the $n\times n$ matrix $A(\cdot)$, $n\times m$ matrix $B(\cdot)$, $n\times n$ matrix $Q(\cdot)$ and $m\times m$ matrix $R(\cdot)$ are of class C^1 on $[t_0, t_1]$. In addition, the matrix $Q(t)$ is positive semidefinite and the matrix $R(t)$ is positive definite for $t\in[t_0, t_1]$.

Let us turn first to the necessary conditions of the maximum principle, Theorem 11.1, as modified for nonautonomous systems in Section 13.6. The function $\hat{H}(\cdot)$ is given by

$$\hat{H}(\lambda, y, t, u) = \tfrac{1}{2}\lambda_0\left[x^T Q(t)x + u^T R(t)u\right] + \hat{\lambda}^T\left[A(t)x + B(t)u\right] \qquad (16.11)$$

where

$$\hat{\lambda} \triangleq [\lambda_1\ \lambda_2 \cdots \lambda_n]^T.$$

Thus the adjoint equations are

$$\dot{\hat{\lambda}}(t) = -\lambda_0(t)Q(t)x(t) - A^T(t)\hat{\lambda}(t) \qquad (16.12)$$

where, by the condition (c) of Theorem 11.1,

$$\lambda_0(t) \equiv \text{constant} \leq 0.$$

Since the control is unconstrained, that is, $U = R^m$, the condition (a) of Theorem 11.1 implies that an extremal control $u(\cdot)$, generating the solution $x(\cdot)$, must satisfy

$$\frac{\partial \hat{H}[\lambda(t), y(t), t, u(t)]}{\partial u_i} = 0, \qquad i = 1, 2, \ldots, m$$

so that

$$\lambda_0(t)R(t)u(t) + B^T(t)\hat{\lambda}(t) = 0. \qquad (16.13)$$

In order to permit the solution of (16.13) for $u(t)$, we need $\lambda_0(t) \neq 0$; we may take $\lambda_0(t) \equiv -1$. Also, since $R(t)$ is positive definite, its inverse exists.

Consequently it follows from (16.13) that

$$u(t) = R^{-1}(t) B^T(t) \hat{\lambda}(t). \tag{16.14}$$

On substitution of (16.14) in (16.9) we obtain

$$\dot{x}(t) = A(t) x(t) + B(t) R^{-1}(t) B^T(t) \hat{\lambda}(t). \tag{16.15}$$

Now, writing (16.15) and (16.12) as a system of order $2n$, we have

$$\begin{bmatrix} \dot{x}(t) \\ \dot{\hat{\lambda}}(t) \end{bmatrix} = \begin{bmatrix} A(t) & B(t) R^{-1}(t) B^T(t) \\ Q(t) & -A^T(t) \end{bmatrix} \begin{bmatrix} x(t) \\ \hat{\lambda}(t) \end{bmatrix}. \tag{16.16}$$

Since (16.16) is of order $2n$ we require $2n$ end conditions. These are $x(t_0) = x^0$ and the consequence of the terminal transversality condition, namely,

$$\hat{\lambda}(t_1) = 0. \tag{16.17}$$

Let $S(t; t_0)$ denote the fundamental matrix (for instance, see Refs. 16.2 and 16.3) of (16.16); that is,

$$\begin{bmatrix} x(t) \\ \hat{\lambda}(t) \end{bmatrix} = S(t; t_0) \begin{bmatrix} x(t_0) \\ \hat{\lambda}(t_0) \end{bmatrix} \tag{16.18}$$

and

$$\begin{bmatrix} x(t_1) \\ \hat{\lambda}(t_1) \end{bmatrix} = S(t_1; t) \begin{bmatrix} x(t) \\ \hat{\lambda}(t) \end{bmatrix}. \tag{16.19}$$

If we partition $S(t_1; t)$ as

$$S(t_1; t) = \begin{bmatrix} S_{11}(t_1; t) & S_{12}(t_1; t) \\ S_{21}(t_1; t) & S_{22}(t_1; t) \end{bmatrix}$$

then, utilizing (16.17), we obtain

$$x(t_1) = S_{11}(t_1; t) x(t) + S_{12}(t_1; t) \hat{\lambda}(t),$$
$$0 = S_{21}(t_1; t) x(t) + S_{22}(t_1; t) \hat{\lambda}(t) \tag{16.20}$$

whence, provided $S_{22}^{-1}(t_1;t)$ exists, we have

$$\hat{\lambda}(t) = -S_{22}^{-1}(t_1;t)S_{21}(t_1;t)x(t). \tag{16.21}$$

Now let us note that

$$\begin{aligned}S(t_1;t_1) &= I_{2n}, \\ S_{11}(t_1;t_1) &= S_{22}(t_1;t_1) = I_n, \\ S_{12}(t_1;t_1) &= S_{21}(t_1;t_1) = 0\end{aligned} \tag{16.22}$$

where I_n denotes the $n \times n$ unitary matrix. Thus at least $S_{22}^{-1}(t_1;t_1)$ exists. In fact, it can be shown that $S_{22}(t_1;t)$ is nonsingular for all $t \in [t_0, t_1]$; see Ref. 16.4.

Let

$$P(t) \triangleq S_{22}^{-1}(t_1;t)S_{21}(t_1;t) \tag{16.23}$$

where, as a consequence of (16.22),

$$P(t_1) = 0. \tag{16.24}$$

Now we rewrite (16.21) as

$$\hat{\lambda}(t) = -P(t)x(t), \tag{16.25}$$

recalling that we require $\hat{\lambda}(t)$ in order to determine $u(t)$ by (16.14). Thus we need $P(t)$. Differentiation of (16.25) results in

$$-\dot{\hat{\lambda}}(t) = \dot{P}(t)x(t) + P(t)\dot{x}(t). \tag{16.26}$$

Now (16.26), together with (16.16) and (16.25), leads at once to

$$-\dot{\hat{\lambda}}(t) = \left[\dot{P}(t) + P(t)A(t) - P(t)B(t)R^{-1}(t)B^T(t)P(t)\right]x(t)$$

and

$$\dot{\hat{\lambda}}(t) = \left[Q(t) + A^T(t)P(t)\right]x(t),$$

or

$$\left[\dot{P}(t) + P(t)A(t) + A^T(t)P(t) - P(t)B(t)R^{-1}(t)B^T(t)P(t) + Q(t)\right]x(t) = 0.$$

This equation must hold for any x^0 at any $t_0 < t_1$; consequently we must have

$$\dot{P}(t) = -P(t)A(t) - A^T(t)P(t) + P(t)B(t)R^{-1}(t)B^T(t)P(t) - Q(t). \tag{16.27}$$

This then is the differential equation for $P(\cdot)$ subject to the end condition (16.24); this equation, termed a Riccati equation, is a nonlinear first-order differential equation for the $n \times n$ matrix $P(t)$. Taking the transpose of both sides of (16.27), and noting that $Q(t)$ and $B(t)R^{-1}(t)B^T(t)$ are symmetric matrices, results in

$$[\dot{P}(t)]^T = -P^T(t)A(t) - A^T(t)P^T(t) + P^T(t)B(t)R^{-1}(t)B^T(t)P^T(t) - Q(t). \tag{16.28}$$

However, since

$$[\dot{P}(t)]^T = \dot{P}^T(t)$$

and

$$P(t_1) = 0 \Rightarrow P^T(t_1) = 0,$$

it follows that $P(\cdot)$ and $P^T(\cdot)$ are solutions of the same differential equation for the same end condition. However, this differential equation has a unique solution; hence we conclude that

$$P(t) = P^T(t) \quad \forall t \in [t_0, t_1].$$

In other words the $n \times n$ matrix $P(t)$ is symmetric; hence the n^2 first-order differential equations for the elements of $P(t)$, the matrix equation (16.27), reduce to $\frac{1}{2}n(n+1)$ equations.

In order to show that the extremal control $u(\cdot) : [t_0, t_1] \to R^m$ given by

$$u(t) = -R^{-1}(t)B^T(t)P(t)x(t) \tag{16.29}$$

is indeed optimal at x^0 we invoke the sufficient conditions of Theorem 15.1. To do so we require a test function $\mathcal{V}(\cdot)$ and hence the optimal cost function $V^*(\cdot)$. The cost function for the extremal control is a candidate for $V^*(\cdot)$.

Consider

$$\frac{d}{dt}\left[x^T(t)\hat{\lambda}(t)\right] = -\frac{d}{dt}\left[x^T(t)P(t)x(t)\right]$$
$$= -x^T(t)\dot{P}(t)x(t) - 2x^T(t)P(t)\dot{x}(t)$$
$$= x^T(t)\left[P(t)A(t) + A^T(t)P(t)\right.$$
$$\left. - P(t)B(t)R^{-1}(t)B^T(t)P(t) + Q(t)\right]x(t)$$
$$- 2x^T(t)P(t)\left[A(t) - B(t)R^{-1}(t)B^T(t)P(t)\right]x(t)$$
$$= x^T(t)Q(t)x(t) + x^T(t)P(t)B(t)R^{-1}(t)B^T(t)P(t)x(t)$$
(16.30)

where we have made use of (16.25), (16.27), and (16.9) with (16.29), and of the symmetry of $P(t)$. Also, since $R(t)$ is positive definite (symmetric), (16.29) implies that

$$u^T(t) = -x^T(t)P(t)B(t)R^{-1}(t)$$

so that finally (16.30) becomes

$$-\frac{d}{dt}\left[x^T(t)P(t)x(t)\right] = x^T(t)Q(t)x(t) + u^T(t)R(t)u(t),$$

which is the integrand of the cost (16.10). Now, upon integrating and invoking the end condition (16.24), we arrive at

$$\frac{1}{2}x^{0T}P(t_0)x^0 = \frac{1}{2}\int_{t_0}^{t_1}\left[x^T(t)Q(t)x(t) + u^T(t)R(t)u(t)\right]dt. \quad (16.31)$$

Thus we are led to the test function $\mathcal{V}(\cdot)$ such that

$$\mathcal{V}(\tilde{x}) = \tfrac{1}{2}x^T P(t)x, \qquad (16.32)$$

where the augmented state is $\tilde{x} = [x \mid t]^T$. The function $\mathcal{V}(\cdot)$ is of class C^1 on

$$X = \{\tilde{x} \in R^{n+1} | t < t_1\}.$$

The condition (i) of Theorem 15.1 is satisfied since $P(t_1) = 0$. To verify the conditions (ii) and (iii) we require

$$\hat{f}_0(x,t,u)+\operatorname{grad}^T \mathcal{V}(\tilde{x})\hat{f}(x,t,u)$$
$$=\tfrac{1}{2}\{x^TQ(t)x+u^TR(t)u+x^T\dot{P}(t)x+2x^TP(t)[A(t)x+B(t)u]\}$$
$$=\tfrac{1}{2}\{u^TR(t)u+2x^TP(t)B(t)u$$
$$+x^TP(t)B(t)R^{-1}(t)R(t)R^{-1}(t)B^T(t)P(t)x\} \qquad (16.33)$$

where we have invoked (16.27) and $R(t)R^{-1}(t)=I_m$. For the extremal control given by (16.29) we recall that

$$u(t)=-R^{-1}(t)B^T(t)P(t)x(t),$$
$$u^T(t)=-x^T(t)P(t)B(t)R^{-1}(t),$$

so that (16.33) becomes

$$u^T(t)R(t)u(t)-2u^T(t)R(t)u(t)+u^T(t)R(t)u(t)=0$$

and the condition (ii) is met.

To verify the condition (iii) we note that

$$u^T=-x^TP(t)B(t)R^{-1}(t) \qquad (16.34)$$

results in the unique *stationary* value of (16.33). Since $R(t)$ is positive definite, this stationary value is a global minimum; for instance, see Ref. 16.5. We have already shown that a control value (16.34) results in the zero value of (16.33). Thus the condition (iii) is also satisfied, and consequently the extremal control given by (16.29) is indeed optimal at $x^0 \in X=E^*$. Hence there is an *optimal feedback control* $k^*(\cdot): E^* \to R^m$ defined by

$$k^*(\tilde{x})=-R^{-1}(t)B^T(t)P(t)x. \qquad (16.35)$$

16.4. The Existence of Feedback Solutions

In the linear-quadratic problem treated in Section 16.3 the optimal feedback control is a function of t and x; it is continuous and continuously differentiable in both variables. Consequently the solution $\tilde{x}^*(\cdot):[t_0,t_1] \to R^{n+1}$, $\tilde{x}^*(t_0)=\tilde{x}^0$, of the state equation (16.2) for the augmented state

$\tilde{x} = [x \mathrel{\vdots} t]^T$, namely,

$$\dot{x}^*(t) = [A(t) - B(t)R^{-1}(t)B^T(t)P(t)]x^*(t),$$
$$\dot{x}^*_{n+1}(t) = 1$$

exists and is unique. Clearly there is no difficulty in this regard here. However, such is not the case with the optimal feedback controls in the problems of 12.1, 14.6, and 14.10; there the optimal feedback controls are *discontinuous* functions of the state variable, x; consequently the conditions assuring the existence of a solution, let alone its uniqueness, are not satisfied (for instance, see Ref. 16.2). As we shall see, this difficulty may be of more than purely theoretical importance; that is, not being able to satisfy the conditions which assure the existence of a solution generated by an optimal feedback control may be indicative of practical difficulties associated with the implementation of the feedback control in the actual system. Our discussion will be brief and largely heuristic and is intended to point to potential problems in the use of discontinuous feedback control; more complete treatments may be found in Refs. 16.6–16.8.

To illustrate our concerns, recall the time-optimal regulator of Section 14.6. The optimal feedback control $k^*(\cdot): R^2 \setminus \{0\} \to R^1$ is defined by

$$k^*(x) = \begin{cases} 1 & \text{for } x \in \mathcal{S}_+, \\ -1 & \text{for } x \in \mathcal{S}_-, \\ -\text{sgn } x_2 & \text{for } x \in \mathcal{S}, \end{cases} \quad (16.36)$$

where

$$\mathcal{S}_+ \triangleq \{x \in R^2 | x_1 < -\tfrac{1}{2}x_2|x_2|\},$$
$$\mathcal{S}_- \triangleq \{x \in R^2 | x_1 > -\tfrac{1}{2}x_2|x_2|\},$$
$$\mathcal{S} \triangleq \{x \in R^2 | x_1 = -\tfrac{1}{2}x_2|x_2|, x_2 \neq 0\}.$$

Thus $k^*(\cdot)$ is discontinuous at $x \in \mathcal{S}$. This does not seem to cause any difficulty with generating a unique solution from any given initial state, x^0, as we saw in Section 14.6. The mathematical model of the *ideal* system is defined by the state equations (14.16) with the control given by (16.36). If there are no perturbations or measurement errors—that is, if the actual system behaves perfectly—then the control value switches at the instant the

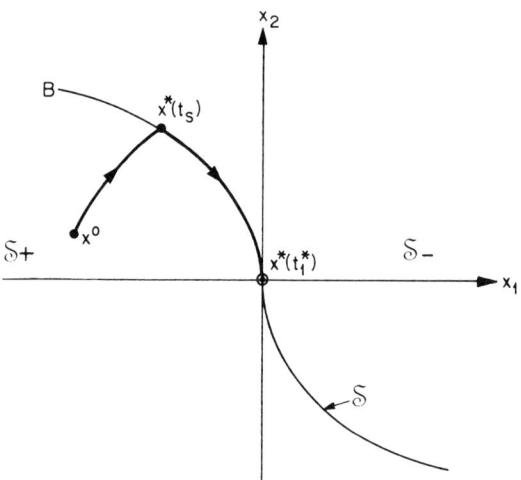

Figure 16.3. An ideal trajectory.

state enters the switching curve, S; thereafter no further switch takes place. This is shown in Figure 16.3. The solution of the state equations with the control given by (16.36) is defined and unique. For instance, if $x^0 \in S_+$, then

$$x_1^*(t) = x_1^0 + x_2^0(t-t_0) + \tfrac{1}{2}(t-t_0)^2,$$
$$x_2^*(t) = x_2^0 + t - t_0$$

for $t \in [t_0, t_s]$, and

$$x_1^*(t) = x_1^*(t_s) + x_2^*(t_s)(t-t_s) - \tfrac{1}{2}(t-t_s)^2,$$
$$x_2^*(t) = x_2^*(t_s) - (t-t_s)$$

for $t \in [t_s, t_1^*]$, where

$$t_s = t_0 + \left[\tfrac{1}{2}(x_2^0)^2 - x_1^0\right]^{1/2} - x_2^0.$$

However, suppose that the actual system is not perfect and that there is an error, possibly "small," in the measurement of the state. For instance, when the actual state reaches S, the measured state is still in S_+ so that no switch is initiated and the trajectory enters S_-. Then, depending on the measurement error, one realizes eventually that the measured state is in S_- and a

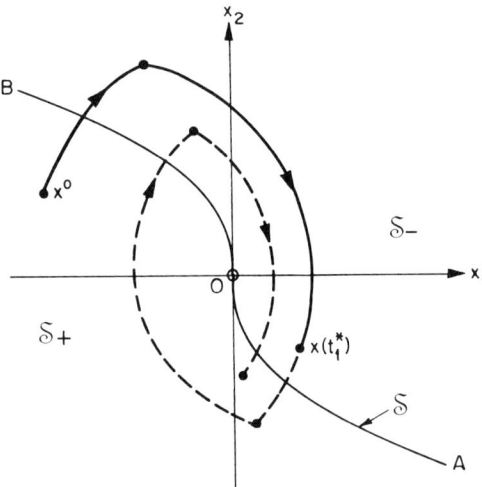

Figure 16.4. A perturbed trajectory.

switch is initiated. The next switch, if any, is implemented after the measured state passes $\mathcal{S}$ again and the trajectory enters $\mathcal{S}_+$, and so on. This is illustrated in Figure 16.4. However, we note that the perturbed trajectory approaches the ideal one as the measurement error diminishes. In particular, if $x(\cdot):[t_0, t_1^*] \to R^n$, $x(t_0) = x^0$, denotes a solution generated by the control for which the switch is delayed, as it might be in the actual system due to measurement error or any other reason, then $x(t) \to x^*(t)$ as the delay in switching approaches zero. Here then we have a situation in which the discontinuity of the optimal feedback control, and hence of the right-hand side of the state equations, with respect to the state variable does not introduce serious difficulties in theory or in practice. This is not always so, as we shall point out in the next section.

16.5. An Illustrative Example

Consider the state equations

$$\dot{x}_1(t) = u_1(t), \qquad \dot{x}_2(t) = u_2(t) \tag{16.37}$$

with the control subject to the constraint

$$u(t) \in U = \{u \in R^2 | 0 \le |u_1| + |u_2| \le 1\}. \tag{16.38}$$

It is desired to obtain a "time-optimal" feedback control for transfer from $x^0 \in R^2$ to $x^1 = 0$.

We begin with the necessary conditions of the maximum principle in order to deduce extremal controls, that is, candidates for open-loop controls which are optimal at x^0. The function $H(\cdot)$ is given by

$$H(\lambda, y, u) = \lambda_0 + \lambda_1 u_1 + \lambda_2 u_2$$

so that, as a consequence of the adjoint equations we have

$$\lambda_i(t) \equiv \text{constant} \triangleq \lambda_i, \qquad i = 0, 1, 2.$$

On selecting $\lambda_0(t) \equiv -1$, satisfying the condition (c) of Theorem 11.1, we see that an extremal control $u(\cdot): [t_0, t_1] \to R^2$ must be such that

$$\max_{u \in U} (-1 + \lambda_1 u_1 + \lambda_2 u_2) = -1 + \lambda_1 u_1(t) + \lambda_2 u_2(t) = 0, \qquad (16.39)$$

satisfying the conditions (a) and (b). Now it is readily verified that the condition (a) is fulfilled for the following four adjoint solutions:

(i) $\lambda_1 = \lambda_2 = 1$,
(ii) $\lambda_1 = \lambda_2 = -1$,
(iii) $\lambda_1 = 1, \quad \lambda_2 = -1$,
(iv) $\lambda_1 = -1, \quad \lambda_2 = 1$.

The extremal controls corresponding to the adjoint solutions (i)–(iv) are such that

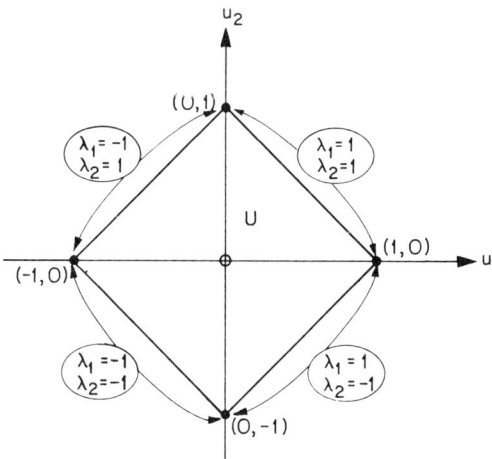

Figure 16.5. Extremal control values.

the condition (b) is satisfied:

(i) $u_1(t)=0$, $u_2(t)=1$ or $u_1(t)=1$, $u_2(t)=0$.
(ii) $u_1(t)=-1$, $u_2(t)=0$ or $u_1(t)=0$, $u_2(t)=-1$.
(iii) $u_1(t)=0$, $u_2(t)=-1$ or $u_1(t)=1$, $u_2(t)=0$.
(iv) $u_1(t)=-1$, $u_2(t)=0$ or $u_1(t)=0$, $u_2(t)=1$.

This is illustrated in Figure 16.5. In other words, an extremal control can switch between the indicated values. To prove that a particular extremal control is optimal at an initial state one can utilize a sufficiency theorem of the field type such as Theorem 15.1 or 15.2. In order to employ a field theorem one requires an extremal control for every initial state in $R^2\setminus\{0\}$ in order to construct a test function $\mathcal{V}(\cdot): R^2\setminus\{0\} \to R^1$.

Let us consider two families of extremal controls.

Case 1.

$$\begin{aligned}
u_1(t)=-1, \quad u_2(t)=0 & \quad \text{if } x_1(t)>0, \; x_2(t)\geq 0, \\
u_1(t)=0, \quad u_2(t)=-1 & \quad \text{if } x_1(t)\leq 0, \; x_2(t)>0, \\
u_1(t)=1, \quad u_2(t)=0 & \quad \text{if } x_1(t)<0, \; x_2(t)\leq 0, \\
u_1(t)=0, \quad u_2(t)=1 & \quad \text{if } x_1(t)\geq 0, \; x_2(t)<0.
\end{aligned} \quad (16.40)$$

Case 2.

$$\begin{aligned}
u_1(t)=0, \quad u_2(t)=-1 & \quad \text{if } x_1(t)\in R^1, \; x_2(t)>0, \\
u_1(t)=-1, \quad u_2(t)=0 & \quad \text{if } x_1(t)>0, \; x_2(t)=0, \\
u_1(t)=1, \quad u_2(t)=0 & \quad \text{if } x_1(t)<0, \; x_2(t)=0, \\
u_1(t)=0, \quad u_2(t)=1 & \quad \text{if } x_1(t)\in R^1, \; x_2(t)<0.
\end{aligned} \quad (16.41)$$

Figures 16.6 and 16.7 show typical trajectories corresponding to Cases 1 and 2, respectively.

For example, in Case 1 the cost for a given initial state, x^0, is obtained by the integration of the state equations as

$$t_1 - t_0 = |x_1^0| + |x_2^0|.$$

Thus we are led to the test function $\mathcal{V}(\cdot)$ such that

$$\mathcal{V}(x) = |x_1| + |x_2|, \quad x \in R^2\setminus\{0\}.$$

Since

$$\operatorname{grad} \mathcal{V}(x) = \left[\frac{x_1}{|x_1|} \quad \frac{x_2}{|x_2|} \right]^T$$

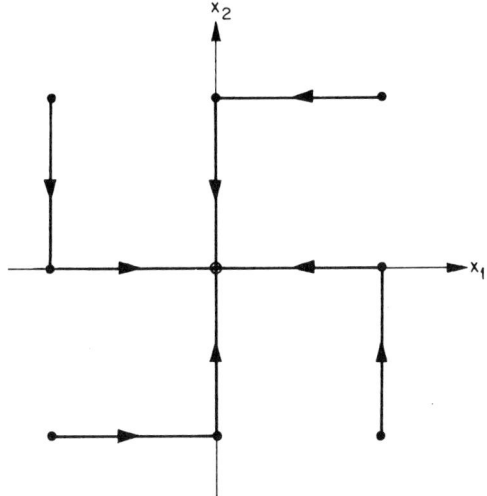

Figure 16.6. Case 1.

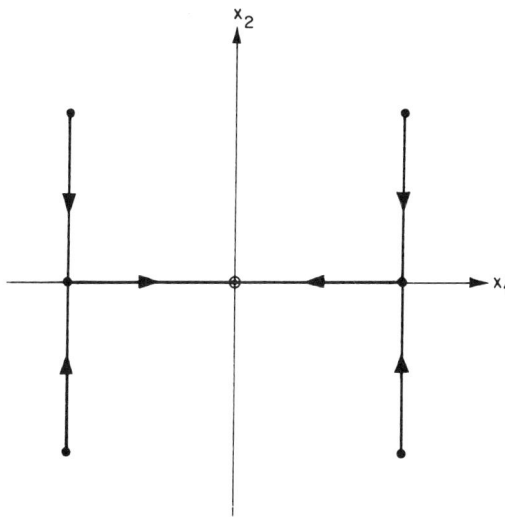

Figure 16.7. Case 2.

we see that grad $\mathcal{V}(x)$ is not defined for

$$x \in \{x \in R^2 \mid x_1 = 0 \text{ or } x_2 = 0\}.$$

Consequently Theorem 15.1 is not applicable. We leave it as an exercise for the reader to construct a finite decomposition and to use Theorem 15.2 in order to prove the optimality of the extremal controls corresponding to Cases 1 and 2, respectively (Exercise 16.5).

Once the optimality of the extremal controls is established, the synthesis of the corresponding optimal feedback controls, $k^*(\cdot)$ and $k^{**}(\cdot)$, follows from (16.40) and (16.41), respectively.

Case 1.

$$\begin{aligned}
k_1^*(x) &= -1, & k_2^*(x) &= 0 & &\text{for } x_1 > 0, \; x_2 \geq 0, \\
k_1^*(x) &= 0, & k_2^*(x) &= -1 & &\text{for } x_1 \leq 0, \; x_2 > 0, \\
k_1^*(x) &= 1, & k_2^*(x) &= 0 & &\text{for } x_1 < 0, \; x_2 \leq 0, \\
k_1^*(x) &= 0, & k_2^*(x) &= 1 & &\text{for } x_1 \geq 0, \; x_2 < 0.
\end{aligned}$$

Case 2.

$$\begin{aligned}
k_1^{**}(x) &= 0, & k_2^{**}(x) &= -1 & &\text{for } x_1 \in R^1, \; x_2 > 0, \\
k_1^{**}(x) &= -1, & k_2^{**}(x) &= 0 & &\text{for } x_1 > 0, \; x_2 = 0, \\
k_1^{**}(x) &= 1, & k_2^{**}(x) &= 0 & &\text{for } x_1 < 0, \; x_2 = 0, \\
k_1^{**}(x) &= 0, & k_2^{**}(x) &= 1 & &\text{for } x_1 \in R^1, \; x_2 < 0.
\end{aligned}$$

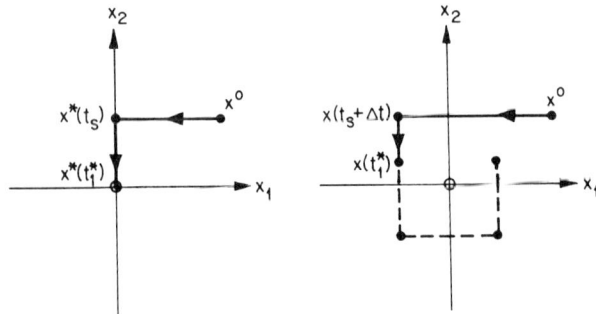

Figure 16.8. Case 1.

Chap. 16 • Feedback Control

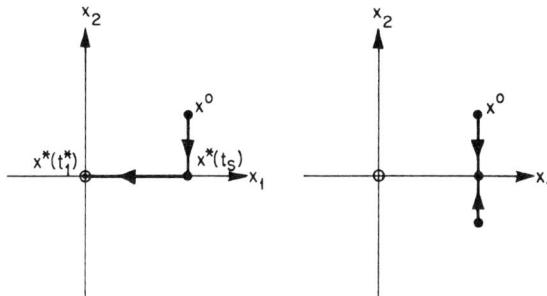

Figure 16.9. Case 2.

Since $k^*(\cdot)$ and $k^{**}(\cdot)$ are discontinuous functions, the usual conditions assuring the existence of solutions of the state equations with optimal feedback control are not satisfied. Nonetheless, as in the time-optimal regulator problem of Section 14.6, the optimal feedback controls $k^*(\cdot)$ and $k^{**}(\cdot)$ generate unique solutions for given initial states. Typical solutions are illustrated in Figures 16.6 and 16.7, respectively. Let us remember, however, that these are solutions of the state equations which model the *ideal* system. Thus, as in Section 16.3, we are concerned with the behavior of an imperfect system, that is, with perturbed solutions. Suppose again that there is a delay, Δt, in effecting a switch in the control resulting in a perturbed solution $x(\cdot):[t_0, t_1^*] \to R^2$, $x(t_0)=x^0$. In Case 1, with the control $k^*(\cdot)$, the resulting perturbed trajectory remains "near" the ideal one; that is, $x(t) \to x^*(t)$ as the delay goes to zero. This is illustrated in Figure 16.8. In Case 2, with the control $k^{**}(\cdot)$, a delay in bringing about a control switch results in a perturbed trajectory that does not stay "near" the ideal one. In particular such a delay leads to the oscillation or "chattering" of the state about the ideal switching state, $x^*(t_s)$. This is so no matter how small the delay is, and consequently $x(t)$ does not approach $x^*(t)$ as the delay goes to zero. This is illustrated in Figure 16.9. In other words, a delay in switching, no matter how small, results in the state "becoming stuck" near the ideal switching state $x^*(t_s)$. This is surely not tolerable. We conclude that the ideally optimal feedback control $k^{**}(\cdot)$ is not suitable in practice.

Exercises

16.1. Recall the time-optimal rocket problem of Section 12.1. Construct an optimal feedback control. Is it unique?

16.2. Recall the time-optimal navigation problem of Section 12.2. Deduce an optimal feedback control by (a) noting the explicit dependence of the optimal open-loop control on the initial state, and (b) utilizing the Hamilton–Jacobi–Bellman equation.

16.3. Recall the problem of Section 13.8. Use the Hamilton–Jacobi–Bellman equation to construct an optimal feedback control.

16.4. Employ the direct sufficiency theorem, Theorem 15.3, to prove the optimality of the extremal control (16.29) for the linear-quadratic problem of Section 16.3.

16.5. Recall the example of Section 16.5. Use Theorem 15.2 to show that the extremal controls satisfying (16.40) and (16.41), respectively, are indeed optimal at $x^0 \in R^2 \setminus \{0\}$.

17

Optimization with Vector – Valued Cost

17.1. Introduction

Up till now we have considered problems involving a *single* performance index or cost, namely, a *scalar-valued* functional. This is predicated on the supposition that one can select a single, overriding criterion in the form of a functional whose value is to be extremized. However, as we have observed a long time ago (Ref. 17.1), "one person's optimum may well be another's pessimum." For example, in designing a product the engineers may wish to maximize strength or durability, the environmentalists may strive for minimum pollution, while the managers may desire to minimize cost or to maximize profit, and so on. In general these goals are not compatible; that is, one would not expect to be able to arrive at a design that results in the simultaneous satisfaction of all parties who can influence the design.

In this section we consider a class of problems that differs only in one respect from the class of problems defined in Chapter 10; namely, in place of a single cost functional $V[x^0, u(\cdot), x(\cdot)]$ we have a finite number of cost functionals $V_i[x^0, u(\cdot), x(\cdot)]$, $i = 1, 2, \ldots, k < \infty$. Since there may not exist a control $u^*(\cdot)$ that renders the minimum value of *each* cost, we are confronted with a dilemma. What is the meaning of optimality in the event of a *vector-valued* cost? We delay answering this question and consider instead a simpler problem—one with performance indices which are functions rather than functionals—in order to arrive at some notion of "optimality" for the case of a vector-valued performance index.

17.2. How to Choose a Cheese

We are planning a party and wish to serve a Camembert for the cheese course. Two kinds of Camembert are available at the local store: an American (domestic) Camembert, A, at $2.00 per pound and a French (imported) Camembert, F, at $3.00 per pound.† Our budget permits a maximum expenditure of $4.00 for the cheese course. Now we are faced with a decision-making process: "How much of each kind of Camembert shall we purchase?" In other words, a decision $d \triangleq [d_1 \ d_2]^T$ consists of two elements, d_1 and d_2, where d_1 is the number of pounds of A and d_2 is the number of pounds of F. Since we can spend from $0.00 to $4.00, the set of decisions permitted to us—if you wish, the decision constraint set—is the set D shown in Figure 17.1. We may make any decision $d \in D$. But which of these is the "best" decision? At this stage of the process that is not a meaningful question. To lend it meaning we must introduce a method of *comparing* decisions.

Suppose first that *taste* (clearly a subjective criterion) is of *sole* importance and that we prefer F two-to-one to A; that is, we are willing to trade one pound of F for two pounds of A, and conversely. Following the example of the economist Edgeworth (Ref. 17.2) we introduce so-called *indifference curves* in the decision space, R^2; these are curves such that any two decisions belonging to a given indifference curve are "equally good" or "equivalent" (in terms of our adopted standard of judgment, here taste), so that we are indifferent about selecting between them. For us such a

†All prices in 1980 dollars.

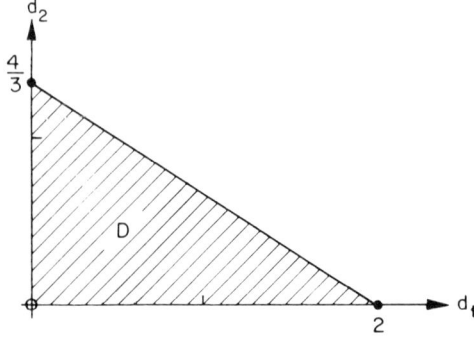

Figure 17.1. Decision constraint set.

Chap. 17 • Optimization with Vector-Valued Cost

taste-indifference curve has the equation

$$d_1 + 2d_2 = c = \text{constant}.$$

Some members of the one-parameter family of taste-indifference curves are shown in Figure 17.2. For example, consider the decisions

$$d^1 = [0 \ 1]^T, \quad d^2 = [1 \ \tfrac{1}{2}]^T, \quad d^3 = [2 \ 0]^T.$$

Clearly all three decisions belong to the same indifference curve, namely, the one for which $c = 2$. Consequently these decisions are equivalent (in terms of taste). We write

$$d^i \sim d^j, \quad i, j \in \{1, 2, 3\}$$

where the symbol $\sim$ stands for "is equivalent to."

Of course we are concerned with the greatest *taste satisfaction*; that is, a decision d^4 lying on a lower curve ($c < 2$) is less *desirable*. Thus we write

$$d^i \succ d^4, \quad i = 1, 2, 3$$

where the symbol $\succ$ stands for "is preferred to." In general if the decisions $\hat{d}$ and d are such that

$$\hat{d}_1 + 2\hat{d}_2 = \hat{c}, \quad d_1 + 2d_2 = c,$$

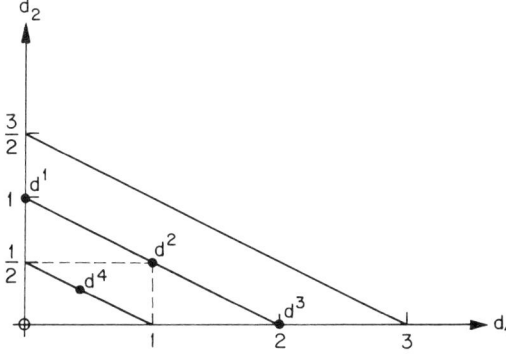

Figure 17.2. Taste-indifference curves.

and we define

$$\hat{D} \triangleq \{d \in R^2 | c \leq \hat{c}\},$$

then

$$d \preccurlyeq \hat{d} \quad \forall d \in \hat{D}$$

where the symbol $\preccurlyeq$ stands for "is not preferred to." This notion of *preference* leads us to an attempt to *order* decisions *numerically*; for a fuller discussion of preference and ordering see Ref. 17.3. If we retain our earlier convention of always *minimizing*, then we introduce a function $\mu_t(\cdot): R^2 \to R^1$ such that

$$\mu_t(d) = -d_1 - 2d_2$$

so that

$$\mu_t(\hat{d}) \leq \mu_t(d) \Leftrightarrow d \preccurlyeq \hat{d}.$$

In this manner we convert the problem of making the "best decision with respect to taste satisfaction" into that of minimizing $\mu_t(\cdot)$ over D; that is, the decision d^t is a best decision with respect to taste satisfaction, or it is *taste-optimal*, if and only if

$$\mu_t(d^t) \leq \mu_t(d) \quad \forall d \in D,$$

As illustrated in Figure 17.3, if taste satisfaction is our sole concern, then our decision must be to purchase only F using all available funds; namely, we arrive at

$$d^t = \begin{bmatrix} 0 & \frac{4}{3} \end{bmatrix}^T.$$

Alas, life is not so uncomplicated and taste is not our only concern. There is also the price of purchasing the cheese. Let us suppose for the moment that the price is our *only* criterion. If we proceed as we did with taste satisfaction, then an indifference curve for price has the equation

$$2d_1 + 3d_2 = c$$

where c is now the total price. Thus, if we select a function $\mu_p(\cdot): R^2 \to R^1$

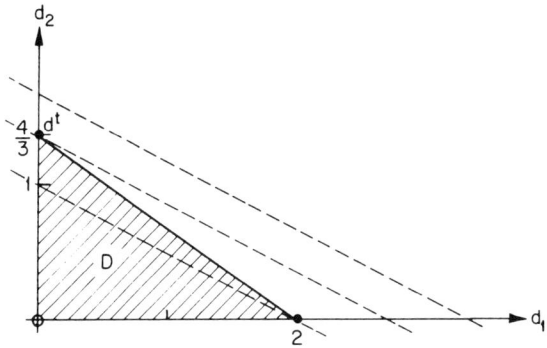

Figure 17.3. The taste-optimal decision.

such that

$$\mu_p(d) = 2d_1 + 3d_2,$$

then

$$\mu_p(\hat{d}) \leq \mu_p(d) \Leftrightarrow d \leq \hat{d}.$$

In other words the decision d^p is a best decision with respect to price, or it is *price-optimal*, if and only if

$$\mu_p(d^p) \leq \mu_p(d) \quad \forall d \in D.$$

As illustrated in Figure 17.4, the price-optimal decision is $d^p = [0\ 0]^T$. This is exactly as expected; if price is our only concern, then buying no cheese results in the minimum purchase price.

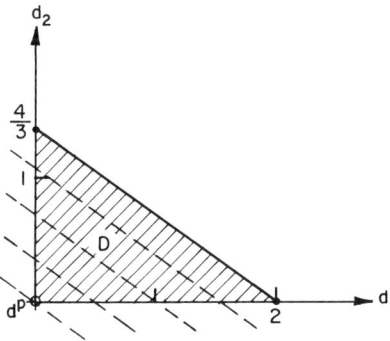

Figure 17.4. The price-optimal decision.

What we actually desire is to reach two goals; we wish to maximize taste satisfaction *and* to minimize price. Clearly, since $d^t \neq d^p$, this cannot be accomplished. There is no decision $d \in D$ which renders the minimum of $\mu_t(\cdot)$ *and* of $\mu_p(\cdot)$. What then is "best" or "optimal" with respect to *both* criteria? To lend more meaning to this question let us construct the set of all vector-valued criterion values, or outcomes, $\mu = [\mu_t \ \mu_p]^T$ which are due to decisions permitted to us, namely, the set

$$M \triangleq \left\{ \mu \in R^2 \mid \mu = \mu(d) = [\mu_t(d) \ \mu_p(d)]^T, d \in D \right\}.$$

This is done easily for our problem and M is shown in Figure 17.5. The decision, d^u, which minimizes *both* $\mu_t(\cdot)$ and $\mu_p(\cdot)$ is not permitted to us; that is, $d^u \notin D$. Hence we refer to it as a "utopian decision."

Upon reflecting on the set of outcomes, M, we notice the following feature. If we select a decision $d^* \in D$ such that $\mu(d^*)$ belongs to the boundary of M and in particular to the face denoted by AB in Figure 17.5, then making any other decision $d \in D$ has one of two effects: either

(i) $\mu_t(d) > \mu_t(d^*)$ or $\mu_p(d) > \mu_p(d^*)$

or

(ii) $\mu_t(d) > \mu_t(d^*)$ and $\mu_p(d) > \mu_p(d^*)$.

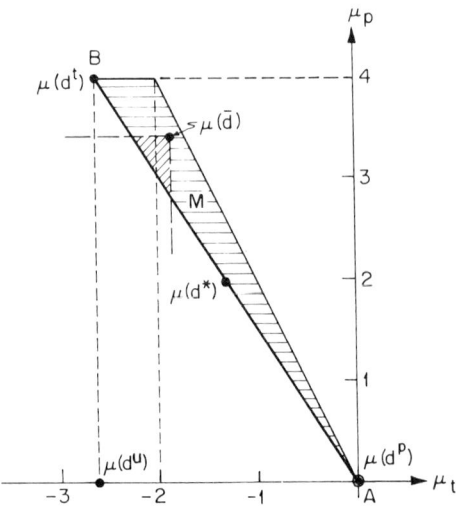

Figure 17.5. The outcome set.

In other words by moving away from d^* we *increase at least one cost* and possibly both, although if one is increased the other might be decreased. All other things being equal we have no reason for selecting d over d^*. Indeed in our problem the set M is such that a decision $\bar{d} \in D$ with $\mu(\bar{d}) \notin AB$ is undesirable, or *inferior*, because there exists a decision $d \in D$ such that

$$\mu_t(d) < \mu_t(\bar{d}) \quad \text{and} \quad \mu_p(d) \leq \mu_p(\bar{d})$$

or

$$\mu_t(d) \leq \mu_t(\bar{d}) \quad \text{and} \quad \mu_p(d) < \mu_p(\bar{d}).$$

The only set of decisions for which this is not the case are those leading to outcomes on the face AB; this is illustrated in Figure 17.5.

Finally, before leaving our particular problem, we determine the set of decisions which result in outcomes on the face AB of M. They are readily found and are shown in Figure 17.6; that is, if $d \in A'B'$ then $\mu(d) \in AB$. Without introducing further considerations we have no way of determining which decision $d \in A'B'$ we ought to make; all we can conclude is that we should *not* make a decision $\bar{d} \notin A'B'$. In other words we have succeeded in reducing the set of decisions of interest to us.

Considerations of the kind discussed in our problem led the Italian engineer and economist Pareto (Ref. 17.4) to define the following concept of "optimality" for decision-making in problems with vector-valued criteria. Given a set of decisions, D, and a set of cost functions (criteria or performance indices)

$$J_i(\cdot) \colon D \to R^1, \qquad i = 1, 2, \ldots, k,$$

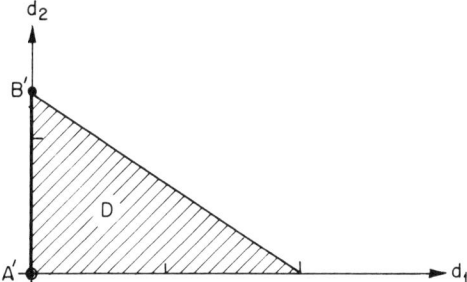

Figure 17.6. Decisions resulting in outcomes on AB.

the decision $d^* \in D$ is *Pareto-optimal* (also sometimes called *efficient*) if and only if for any $d \in D$, $d \neq d^*$,

$$J_i(d) = J_i(d^*) \qquad \forall i \in \{1, 2, \ldots, k\}$$

or there is at least one $i \in \{1, 2, \ldots, k\}$ such that

$$J_i(d) > J_i(d^*).$$

There are other definitions of Pareto-optimality which are equivalent to the one given here (see Refs. 17.5 and 17.6). Another definition in common use is the following. Given a set of decisions, D, and a set of cost functions

$$J_i(\cdot) : D \to R^1, \qquad i = 1, 2, \ldots, k,$$

the decision $d^* \in D$ is *Pareto-optimal* if and only if for all $d \in D$

$$J_i(d) \leq J_i(d^*) \qquad \forall i \in \{1, 2, \ldots, k\}$$

implies that

$$J_i(d) = J_i(d^*) \qquad \forall i \in \{1, 2, \ldots, k\}.$$

We leave it as an exercise for the reader to prove the equivalence of the two definitions of Pareto-optimality given above (Exercise 17.1).

Having provided some motivation for the concept of Pareto-optimality, we turn now to the application of this idea to systems of the kind introduced in Chapter 10 and modified to allow for a vector-valued cost.

17.3. Pareto-Optimal Control

Consider the system defined in Chapter 10, namely, one with the state equation (10.2) and the target set θ^1. The definitions of an "admissible control" and of a "control that is feasible at x^0" are also unaltered. However, in place of the scalar-valued cost (10.4), we have now a vector-valued cost with the components

$$V_i[x^0, u(\cdot), x(\cdot)] \triangleq \int_{t_0}^{t_1} f_0^i[x(t), u(t)] \, dt, \qquad i = 1, 2, \ldots, k, \quad (17.1)$$

where the functions $f_0^i(\cdot): R^n \times R^m \to R^1$ are of the same class as the function $f_0(\cdot)$.

In conformity with the discussion in Section 17.2 we associate a control $u(\cdot)$ with a decision d, the set $\mathcal{U}(x^0)$ of controls which are feasible at x^0 with the decision set D, and a cost component $V_i[x^0, u(\cdot), x(\cdot)]$ with $J_i(d)$. Then, in place of the definition of a "control that is optimal at x^0" we have the following definition. The control $u^*(\cdot): [t_0, t_1^*] \to R^m$, generating the solution $x^*(\cdot): [t_0, t_1^*] \to R^n$ such that $x^*(t_0) = x^0$ and $x^*(t_1^*) \in \theta^1$, is *Pareto-optimal at* x^0 if and only if for all $u(\cdot) \in \mathcal{U}(x^0)$, $u(\cdot) \neq u^*(\cdot)$, that is, for every admissible $u(\cdot): [t_0, t_1] \to R^m$ generating $x(\cdot): [t_0, t_1] \to R^n$, $x(t_0) = x^0$ and $x(t_1) \in \theta^1$, either

$$V_i[x^0, u(\cdot), x(\cdot)] = V_i[x^0, u^*(\cdot), x^*(\cdot)] \qquad \forall i \in \{1, 2, \ldots, k\}$$

or there is at least one $i \in \{1, 2, \ldots, k\}$ such that

$$V_i[x^0, u(\cdot), x(\cdot)] > V_i[x^0, u^*(\cdot), x^*(\cdot)].$$

As in problems with scalar-valued costs, here too we wish to have *necessary* conditions in order to obtain candidates for controls which are Pareto-optimal at x^0, as well as *sufficient* conditions in order to establish that a particular control is indeed Pareto-optimal at x^0.

17.4. Necessary Conditions for Pareto-Optimality

Let us consider first conditions which must be satisfied if a control is Pareto-optimal at x^0. We have the following theorem (for instance, see Ref. 17.7).

Theorem 17.1. *If the control* $u^*(\cdot): [t_0, t_1^*] \to R^m$, *generating the solution* $x^*(\cdot): [t_0, t_1^*] \to R^n$, $x^*(t_0) = x^0$, *is Pareto-optimal at* x^0, *then it is optimal at* x^0 *for the system with the scalar-valued cost*

$$V_i[x^0, u(\cdot), x(\cdot)], \qquad i \in \{1, 2, \ldots, k\}$$

and subject to the isoperimetric constraints

$$V_j[x^0, u(\cdot), x(\cdot)] \leq V_j[x^0, u^*(\cdot), x^*(\cdot)], \qquad j = 1, 2, \ldots, k \text{ and } j \neq i.$$

Proof. Suppose that the theorem is false. Then there is an $i \in \{1, 2, \ldots, k\}$ and a control $u(\cdot) \in \mathcal{U}(x^0)$ with the corresponding solution $x(\cdot)$ such that

$$V_i[x^0, u(\cdot), x(\cdot)] < V_i[x^0, u^*(\cdot), x^*(\cdot)]$$

and

$$V_j[x^0, u(\cdot), x(\cdot)] \leq V_j[x^0, u^*(\cdot), x^*(\cdot)],$$

$j = 1, 2, \ldots, k$ and $j \neq i$. But that contradicts the Pareto-optimality of $u^*(\cdot)$ at x^0. □

Theorem 17.1 is very useful because it implies that the necessary conditions for optimal control subject to isoperimetric constraints, Theorem 13.1, are also necessary for Pareto-optimality in the problem with a vector-valued cost. If we let

$$\alpha \triangleq [\alpha_1 \ \alpha_2 \cdots \alpha_k]^T,$$
$$\hat{\lambda} \triangleq [\lambda_1 \ \lambda_2 \cdots \lambda_n]^T, \quad (17.2)$$
$$\hat{H}(\hat{\lambda}, \alpha, x, u) \triangleq \sum_{i=1}^{k} \alpha_i f_0^i(x, u) + \sum_{i=1}^{n} \lambda_i f_i(x, u),$$

then the following theorem is an immediate consequence of Theorem 17.1 and Theorem 13.1.

Theorem 17.2. *If* $u^*(\cdot): [t_0, t_1^*] \to R^m$, *generating the solution* $x^*(\cdot): [t_0, t_1^*] \to R^n$, $x^*(t_0) = x^0$, *is Pareto-optimal at* x^0, *then there are a solution* $\hat{\lambda}(\cdot): [t_0, t_1^*] \to R^n$ *of* (13.133) *for* $j = 1, 2, \ldots, n$, *and a constant* $\alpha \in R^k$, *where* $[\hat{\lambda}^T(t) \vdots \alpha^T] \neq 0$, *such that*

(a) $\min_{u \in U} \hat{H}[\hat{\lambda}(t), \alpha, x^*(t), u] = \hat{H}[\hat{\lambda}(t), \alpha, x^*(t), u^*(t)],$

(b) $\hat{H}[\hat{\lambda}(t), \alpha, x^*(t), u^*(t)] = 0$

for all $t \in [t_0, t_1^*]$, *holding with* $u^*(\bar{t} - 0)$ *and* $u^*(\bar{t} + 0)$ *if* $u^*(\cdot)$ *is discontinuous at* $\bar{t} \in (t_0, t_1^*)$,

(c) $\alpha_i \geq 0, \quad i = 1, 2, \ldots, k,$

and

(d) *the terminal transversality condition* (13.123)–(13.124) *is satisfied.*

Before presenting an example to illustrate the utilization of Theorem 17.2 we turn to a brief discussion of *sufficient* conditions for Pareto-optimality.

17.5. Sufficient Conditions for Pareto-Optimality

One can readily prove two lemmas which embody conditions whose satisfaction assures that a particular control is Pareto-optimal at x^0 (for instance, see Ref. 17.6).

Lemma 17.1. *The control $u^*(\cdot) \in \mathcal{U}(x^0)$, generating the solution $x^*(\cdot)$, is Pareto-optimal at x^0 if there exists a constant $\alpha \in R^k$ with $\alpha_i > 0$ for $i = 1, 2, \ldots, k$ and $\sum_{i=1}^{k} \alpha_i = 1$, such that*

$$\sum_{i=1}^{k} \alpha_i V_i[x^0, u(\cdot), x(\cdot)] \geq \sum_{i=1}^{k} \alpha_i V_i[x^0, u^*(\cdot), x^*(\cdot)] \quad (17.3)$$

for every $u(\cdot) \in \mathcal{U}(x^0)$ generating the solution $x(\cdot)$.

Proof. Consider a $u(\cdot) \in \mathcal{U}(x^0)$. If the equality holds in (17.3) then either

$$V_i[x^0, u(\cdot), x(\cdot)] = V_i[x^0, u^*(\cdot), x^*(\cdot)] \quad \forall i \in \{1, 2, \ldots, k\}$$

or there exist i and $j \in \{1, 2, \ldots, k\}$, $i \neq j$, such that

$$V_i[x^0, u(\cdot), x(\cdot)] > V_i[x^0, u^*(\cdot), x^*(\cdot)]$$

and

$$V_j[x^0, u(\cdot), x(\cdot)] < V_j[x^0, u^*(\cdot), x^*(\cdot)].$$

If the inequality holds in (17.3) then there is an $i \in \{1, 2, \ldots, k\}$ such that

$$V_i[x^0, u(\cdot), x(\cdot)] > V_i[x^0, u^*(\cdot), x^*(\cdot)].$$

In any event the conditions of the definition of Pareto-optimality at x^0 are fulfilled. $\square$

We leave it as an exercise for the reader to prove the following lemma (Exercise 17.2).

Lemma 17.2. *The control $u^*(\cdot) \in \mathcal{U}(x^0)$, generating the solution $x^*(\cdot)$, is Pareto-optimal at x^0 if there exists a constant $\alpha \in R^k$ with $\alpha_i \geq 0$ for*

$i = 1, 2, \ldots, k$ and $\sum_{i=1}^{k} \alpha_i = 1$, such that

$$\sum_{i=1}^{k} \alpha_i V_i[x^0, u(\cdot), x(\cdot)] > \sum_{i=1}^{k} \alpha_i V_i[x^0, u^*(\cdot), x^*(\cdot)] \quad (17.4)$$

for every $u(\cdot) \in \mathcal{U}(x^0)$ generating the solution $x(\cdot)$, and $u(\cdot) \neq u^*(\cdot)$.

Note that, unlike in Lemma 17.1, not all components of α need to be nonzero; however the minimum of $\sum_{i=1}^{k} \alpha_i V_i[x^0, u(\cdot), x(\cdot)]$ must be achieved by a unique control, $u^*(\cdot)$.

Again, as with the necessary conditions, we have reduced the sufficient conditions for Pareto-optimality to sufficient conditions for optimality with the *scalar-valued* cost

$$\sum_{i=1}^{k} \alpha_i V_i[x^0, u(\cdot), x(\cdot)]. \quad (17.5)$$

If there is an appropriate $\alpha \in R^k$ and if $u^*(\cdot)$ is optimal at x^0 for the problem with the scalar-valued cost (17.5), then $u^*(\cdot)$ is Pareto-optimal at x^0 for the problem with the vector-valued cost.

The use of the direct sufficiency conditions of Theorem 15.3 is illustrated in the example of the next section. Other direct sufficiency conditions are discussed in Ref. 17.8. A sufficiency theorem of the field type is utilized in the problem treated in Ref. 17.6.

17.6. An Illustrative Example

Two divisions of a firm producing competitive products wish to determine their rates of advertising expenditures in order to maximize each division's profit over a season of prescribed length; for instance, see Ref. 17.8. Let t be the time, x_i be the gross revenue rate, c_i be the fraction of revenue after marginal cost, and u_i be the advertising expenditure rate for division $i = 1, 2$. Thus the profit of division i during the time interval $[t_0, t_1]$ is

$$\int_{t_0}^{t_1} [c_i x_i(t) - u_i(t)] \, dt.$$

The ith division's gross revenue rate change depends on its own gross revenue and on its own as well as on its sister division's advertising expenditures. It is assumed that the gross revenue rate decreases if the division does not advertise as well as because of the sister division's advertising, and that advertising tends to

Chap. 17 • Optimization with Vector-Valued Cost

increase a division's gross revenue rate but with diminishing returns. Consequently we adopt the following state equations:

$$\begin{aligned}\dot{x}_1(t) &= a_1 u_1(t) - b_1 u_1^2(t) - x_1(t) - u_2(t), \\ \dot{x}_2(t) &= a_2 u_2(t) - b_2 u_2^2(t) - x_2(t) - u_1(t),\end{aligned} \qquad (17.6)$$

where the a_i and b_i are given positive constants. The initial revenue rates, $x_i(t_0) = x_i^0$, and the season $[t_0, t_1]$, are prescribed.

The advertising expenditure rates must be nonnegative; that is,

$$u_i(t) \geq 0, \quad i = 1, 2. \qquad (17.7)$$

Since the profits are to be maximized, the costs to be minimized are

$$V_i[x^0, u(\cdot), x(\cdot)] = \int_{t_0}^{t_1} [u_i(t) - c_i x_i(t)] \, dt, \quad i = 1, 2. \qquad (17.8)$$

If one supposes that divisions of the *same* firm must act as a coalition, then we have a system with a single control, $u = [u_1 \, u_2]^T$, but with a *vector-valued* cost whose components are defined by (17.8). We are interested in the Pareto-optimal controls. To obtain candidates we invoke Theorem 17.2, keeping in mind that the system is nonautonomous because the interval $[t_0, t_1]$ is prescribed; that is, we consider the augmented state $\tilde{x} \triangleq [x_1 \, x_2 \, t]^T$.

To be specific let us take

$$\begin{aligned} a_1 &= a_2 = 12, \\ b_1 &= b_2 = 2, \\ c_1 &= c_2 = \tfrac{1}{3}, \\ t_0 &= 0, \quad t_1 = 1. \end{aligned}$$

On applying the conditions of Theorem 17.2 (Exercise 17.3) it is readily shown that

$$\begin{aligned} \lambda_1(t) &= \tfrac{1}{3}\alpha_1(e^{t-1} - 1), \\ \lambda_2(t) &= \tfrac{1}{3}\alpha_2(e^{t-1} - 1), \\ \lambda_3(t) &\equiv \text{constant} \triangleq \lambda_3. \end{aligned}$$

Then the condition (a) of Theorem 17.2 implies the following Pareto-optimal candidates: For $i=1, j=2$ and $i=2, j=1$,

$$u_i(t) = \begin{cases} \dfrac{3}{4}(e^{t-1} - 1) + 3 - \dfrac{\alpha_j}{4\alpha_i}, & t \in [0, t_i), \\ 0, & t \in [t_i, 1] \end{cases} \qquad (17.9)$$

where

$$t_i = 1 + \ln\left[1 + 3(\alpha_j/\alpha_i - 12)^{-1}\right]$$

if

$$\alpha_j/\alpha_i < 3(e^{-1} - 1)^{-1} + 12,$$

and

$$u_i(t) \equiv 0 \tag{17.10}$$

if

$$\frac{\alpha_j}{\alpha_i} \geq 3(e^{-1} - 1)^{-1} + 12.$$

In other words there is a two-parameter family of Pareto-optimal candidates. To show that they are indeed Pareto-optimal at $\tilde{x}^0$ we invoke Lemma 17.1 together with the direct sufficiency conditions of Theorem 15.3. Consider $\alpha \in R^2$ such that $\alpha_i > 0$, $\alpha_1 + \alpha_2 = 1$. We must show that the control (17.9)–(17.10) results in the minimum of the corresponding scalar-valued cost (17.5). We employ Theorem 15.3 using the test function $p(\cdot):[t_0, t_1] \to R^2$ with $p_i(t) = \lambda_i(t)$. The condition (i) of Theorem 15.3 is readily verified. After invoking

$$\dot{\lambda}_i(t) = 1 + \lambda_i(t)$$

we check the sign of

$$[\alpha_1 + 12\lambda_1(t) - \lambda_2(t)][u_1 - u_1(t)] - 2\lambda_1(t)[u_1^2 - u_1^2(t)]$$
$$+ [\alpha_2 + 12\lambda_2(t) - \lambda_1(t)][u_2 - u_2(t)] - 2\lambda_2(t)[u_2^2 - u_2^2(t)] \tag{17.11}$$

for all $u_i \geq 0$, $i = 1, 2$, and $t \in [0, 1]$; it is nonnegative since the $u(t) = [u_1(t) \ u_2(t)]^T$ given by (17.9)–(17.10) results in the minimum value of the expression (17.11). The condition (ii) of Theorem 15.3 is satisfied since $\lambda_i(1) = 0$.

Exercises

17.1. Prove that the two definitions of Pareto-optimality stated in Section 17.2 are equivalent.

17.2. Prove Lemma 17.2.

17.3. Utilize Theorem 17.2 in order to deduce candidates for Pareto-optimal controls in the example of Section 17.6.

Chap. 17 • Optimization with Vector-Valued Cost   299

17.4. Consider the state equation

$$\dot{x}_1(t) = -[u_1(t) + u_2(t)]x_1(t)$$

with the control subject to the constraints

$$u_1(t) \in [0, a], \qquad u_2(t) \in [0, b]$$

where a and b are positive constants. It is desired to transfer the state from a given initial one, $x_1(t_0) = x_1^0$, to a given terminal one, $x_1(t_1) = x_1^1$, with $0 < x_1^1 < x_1^0$. The vector-valued cost has the components

$$\int_{t_0}^{t_1} [k_i + x_i(t)u_i(t)]\, dt, \qquad i = 1, 2$$

where the k_i are positive constants. Use Theorem 17.2 to deduce candidates for Pareto-optimal controls. Then employ a sufficiency theorem of the field type to prove that the candidates are indeed Pareto-optimal.

17.5. Consider the state equations

$$\dot{x}_1(t) = u_1(t) + u_2(t),$$
$$\dot{x}_2(t) = u_1^2(t),$$
$$\dot{x}_3(t) = u_2^2(t),$$

with the control subject to the constraints

$$u_i(t) \in [0, 1], \qquad i = 1, 2.$$

The initial state, $x(t_0) = [0\,0\,0]^T$, as well as the interval $[t_0, t_1]$ are specified. The cost components are

$$\int_{t_0}^{t_1} [\dot{x}_2(t) - \dot{x}_1(t)]\, dt,$$

$$\int_{t_0}^{t_1} [\dot{x}_3(t) - \dot{x}_1(t)]\, dt.$$

Use the necessary conditions of Theorem 17.2 in order to deduce candidates for Pareto-optimal controls.

References

1.1. MIELE, A., *Flight Mechanics*-1, Addison-Wesley, Reading, Massachusetts, 1962.

2.1. APOSTOL, T. M., *Calculus*, Blaisdell Publishing Company, Waltham, Massachusetts, 1967.
2.2. BARTLE, R. G., *The Elements of Real Analysis*, John Wiley and Sons, New York, 1966.
2.3. BUCK, R. C., *Advanced Calculus*, McGraw-Hill Book Company, New York, 1978.
2.4. PARS, L. A., *A Treatise on Analytical Dynamics*, Heinemann, London, 1965 (also Ox Bow Press, Woodbridge, Connecticut, 1979).
2.5. ROSENBERG, R. M., *Analytical Dynamics of Discrete Systems*, Plenum Press, New York, 1977.

3.1. ZITRON, N. R., A continuous model of optimal-cost routes, *Journal of Optimization Theory and Applications*, Vol. 14, No. 3, 1974.
3.2. BUCK, R. C., *Advanced Calculus*, McGraw-Hill Book Company, New York, 1978.

4.1. BOYCE, W. E., and DIPRIMA, R. C., *Elementary Differential Equations and Boundary Value Problems*, John Wiley and Sons, New York, 1977.
4.2. PARS, L. A., *A Treatise on Analytical Dynamics*, Heinemann, London, 1965 (also Ox Bow Press, Woodbridge, 1979).
4.3. ROSENBERG, R. M., *Analytical Dynamics of Discrete Systems*, Plenum Press, New York, 1977.
4.4. LEITMANN, G., Some remarks on Hamilton's principle, *Journal of Applied Mechanics*, Vol. 30, No. 1, 1963.

5.1. BARTLE, R. G., *The Elements of Real Analysis*, John Wiley and Sons, New York, 1966.

6.1. KAPLAN, W., *Ordinary Differential Equations*, Addison-Wesley, Reading, Massachusetts, 1961.

8.1. EWING, G. M., *Calculus of Variations with Applications*, W. W. Norton, New York, 1969.
8.2. MCSHANE, E. J., The calculus of variations from the beginning through optimal control theory, *Optimal Control and Differential Equations*, edited by A. B. Schwarzkopf, W. G. Kelley, and S. B. Eliason, Academic Press, New York, 1978.

References

9.1. CANNON, R. H., Jr., *Dynamics of Physical Systems*, McGraw-Hill Book Company, New York, 1967.

10.1. CODDINGTON, E. A., and LEVINSON, N., *Theory of Ordinary Differential Equations*, McGraw-Hill Book Company, New York, 1955.
10.2. LEE, E. B., and MARKUS, L., *Foundations of Optimal Control Theory*, John Wiley and Sons, New York, 1967.
10.3. LEITMANN, G., and STALFORD, H., A note on termination in optimal control problems, *Journal of Optimization Theory and Applications*, Vol. 8, No. 3, 1971.
10.4. BELLMAN, R. E., *Dynamic Programming*, Princeton University Press, Princeton, New Jersey, 1957.
10.5. BELLMAN, R. E., and DREYFUS, S. E., *Applied Dynamic Programming*, Princeton University Press, Princeton, New Jersey, 1962.
10.6. DREYFUS, S. E., and LAW, A. M., *The Art and Theory of Dynamic Programming*, Academic Press, New York, 1977.

11.1. KAPLAN, W., *Ordinary Differential Equations*, Addison-Wesley, Reading, Massachusetts, 1961.
11.2. CHEN, C. -T., *Introduction to Linear System Theory*, Holt, Rinehart and Winston, New York, 1970.
11.3. CODDINGTON, E. A., and LEVINSON, N., *Theory of Ordinary Differential Equations*, McGraw-Hill Book Company, New York, 1955.
11.4. BLAQUIÈRE, A., and LEITMANN, G., On the geometry of optimal processes, *Topics in Optimization*, edited by G. Leitmann, Academic Press, New York, 1967.
11.5. PONTRYAGIN, L. S., BOLTYANSKII, V. G., GAMKRELIDZE, R. V., and MISHCHENKO, E. F., *The Mathematical Theory of Optimal Processes*, Interscience Publishers, New York, 1962.
11.6. ATHANS, M., and FALB, P. L., *Optimal Control*, McGraw-Hill Book Company, New York, 1966.
11.7. BELLMAN, R. E., *Dynamic Programming*, Princeton University Press, Princeton, New Jersey, 1957.
11.8. DREYFUS, S. E., *Dynamic Programming and the Calculus of Variations*, Academic Press, New York, 1965.
11.9. LARSON, R. E., and CASTI, J. L., *Principles of Dynamic Programming*, Marcel Dekker, New York, 1978.
11.10. LEE, E. B., and MARKUS, L., *Foundations of Optimal Control Theory*, John Wiley and Sons, New York, 1967.

12.1. LEITMANN, G., Variational problems with bounded control variables, *Optimization Techniques*, edited by G. Leitmann, Academic Press, New York, 1962.
12.2. ZERMELO, E., Über das Navigationsproblem bei ruhender oder veränderlicher Windverteilung, *Zeitschrift für Angewandte Mathematik und Mechanik*, Vol. 11, No. 2, 1931.
12.3. HESTENES, M. R., *Calculus of Variations and Optimal Control Theory*, John Wiley and Sons, New York, 1966.

13.1. LEITMANN, G., On a class of variational problems in rocket flight, *Journal of the Aerospace Sciences*, Vol. 26, No. 9, 1959.
13.2. ISAEV, V. K., L. S. Pontryagin's maximum principle and optimal programming of rocket thrust, *Automation and Remote Control*, Vol. 22, pp. 881 ff., 1961.
13.3. LEITMANN, G., A note on a class of variational problems in rocket flight, *Journal of the Aeronautical Sciences*, Vol. 29, No. 8, 1962.

References

13.4. APOSTOL, T. M., *Calculus*, Blaisdell, Waltham, Massachusetts, 1967.
13.5. BUCK, R. C., *Advanced Calculus*, McGraw-Hill Book Company, New York, 1978.
13.6. CODDINGTON, E. A., and LEVINSON, N., *Theory of Ordinary Differential Equations*, McGraw-Hill Book Company, New York, 1955.
13.7. SCHMITENDORF, W., Pontryagin's principle for problems with isoperimetric constraints and for problems with inequality terminal constraints, *Journal of Optimization Theory and Applications*, Vol. 18, No. 4, 1976.

14.1. KAPLAN, W., *Ordinary Differential Equations*, Addison-Wesley, Reading, Massachusetts, 1961.
14.2. BELL, D. J., and JACOBSON, D. H., *Singular Optimal Control Problems*, Academic Press, New York, 1975.
14.3. HIBBS, A. R., Optimum burning program for horizontal flight, *ARS Journal*, Vol. 22, No. 4, 1952.
14.4. MIELE, A., The calculus of variations in applied aerodynamics and flight mechanics, *Optimization Techniques*, edited by G. Leitmann, Academic Press, New York, 1962.
14.5. MIELE, A., *Flight Mechanics—1*, Addison-Wesley, Reading, Massachusetts, 1962.

15.1. STALFORD, H., Sufficient conditions for optimal control with state and control constraints, *Journal of Optimization Theory and Applications*, Vol. 7, No. 2, 1971.
15.2. LEITMANN, G., and STALFORD, H., A sufficiency theorem for optimal control, *Journal of Optimization Theory and Applications*, Vol. 8, No. 3, 1971.
15.3. PETERSON, D. W., and ZALKIND, J. H., A review of direct sufficient conditions in optimal control theory, *International Journal on Control*, Vol. 28, No. 4, 1978.
15.4. OSTER, G., and WILSON, E. O., *Caste and Ecology in the Social Insects*, Princeton University Press, Princeton, New Jersey, 1978.
15.5. COHEN, D., Maximizing final yield when growth is limited by time or by limiting resources, *Journal of Theoretical Biology*, Vol. 33, pp. 299–307, 1971.
15.6. VINCENT, T. L., and PULLIAM, H. R., Evolution of life history strategies for an asexual annual plant model, *Theoretical Population Biology*, Vol. 17, pp. 215–231, 1980.
15.7. LUENBERGER, D. G., A nonlinear economic control problem with a linear feedback solution, *IEEE Transactions on Automatic Control*, Vol. AC-20, No. 2, 1975.

16.1. LEITMANN, G., A note on optimal open-loop and closed-loop control, *Journal of Dynamical Systems, Measurement and Control*, Vol. 96, No. 3, 1974.
16.2. CODDINGTON, E. A., and LEVINSON, N., *Theory of Ordinary Differential Equations*, McGraw-Hill Book Company, New York, 1955.
16.3. CHEN, C. -T., *Introduction to Linear System Theory*, Holt, Rinehart and Winston, New York, 1970.
16.4. KALMAN, R. E., Contributions to the theory of optimal control, *Boletin Sociedad Matematica Mexicana*, Serie 2, Vol. 5, No. 1, 1960.
16.5. ROCKAFELLAR, R. T., *Convex Analysis*, Princeton University Press, Princeton, New Jersey, 1970.
16.6. HERMES, H., Discontinuous vector fields and feedback control, *Differential Equations and Dynamical Systems*, edited by J. K. Hale and J. P. LaSalle, Academic Press, New York, 1967.
16.7. AIZERMAN, M. A., and PIANITSKII, E. S., Theory of dynamic systems which incorporate elements with incomplete information and its relation to the theory of discontinuous systems, *Journal of the Franklin Institute*, Vol. 306, No. 6, 1978.
16.8. HAJEK, O., Discontinuous differential equations, *Journal of Differential Equations*, Vol. 32, No. 2, 1979.

- 17.1. LEITMANN, G., editor, *Optimization Techniques*, Academic Press, New York, 1962.
- 17.2. EDGEWORTH, F. Y., *Mathematical Psychics*, Keagan, London, 1881.
- 17.3. STADLER, W., Preference optimality in multicriteria control and programming problems, *Nonlinear Analysis, Theory, Methods and Applications*, Vol. 4, No. 1, 1980.
- 17.4. PARETO, V., *Manual of Political Economy*, translated by A. S. Schwier, MacMillan, New York, 1971.
- 17.5. STARR, A. W., and HO, Y. C., Nonzero-sum differential games, *Journal of Optimization Theory and Applications*, Vol. 3, No. 3, 1969.
- 17.6. LEITMANN, G., *Cooperative and Non-Cooperative Many Player Differential Games*, Springer Verlag, Vienna, 1974.
- 17.7. SCHMITENDORF, W. E., and LEITMANN, G., A simple derivation of necessary conditions for Pareto optimality, *IEEE Transactions on Automatic Control*, Vol. AC-19, No. 5, 1974.
- 17.8. LEITMANN, G., and SCHMITENDORF, W., Some sufficiency conditions for Pareto-optimal control, *Journal of Dynamical Systems, Measurement and Control*, Vol. 95, No. 4, 1973.

Bibliography

AKHIEZER, N. I., *The Calculus of Variations*, Blaisdell Publishing Company, New York, 1962.
ATHANS, M., and FALB, P. L., *Optimal Control*, McGraw-Hill Book Company, New York, 1966.
BALAKRISHNAN, A. V., and NEUSTADT, L. W. (editors), *Computing Methods in Optimization Problems*, Academic Press, New York, 1964.
BELL, D. J., and JACOBSON, D. H., *Singular Optimal Control Problems*, Academic Press, New York, 1975.
BELLMAN, R. E., *Dynamic Programming*, Princeton University Press, Princeton, New Jersey, 1957.
BELLMAN, R. E., *Adaptive Control Processes: A Guided Tour*, Princeton University Press, Princeton, New Jersey, 1961.
BELLMAN, R. E. (editor), *Mathematical Optimization Techniques*, University of California Press, Berkeley, 1963.
BELLMAN, R. E., *Introduction to the Mathematical Theory of Control Processes*, Vol. I, Academic Press, New York, 1967.
BELLMAN, R. E., *Introduction to the Mathematical Theory of Control Processes*, Vol. II, Academic Press, New York, 1971.
BELLMAN, R. E., and DREYFUS, S. E., *Applied Dynamic Programming*, Princeton University Press, Princeton, New Jersey, 1962.
BERKOVITZ, L. D., *Optimal Control Theory*, Springer-Verlag, New York, 1974.
BLISS, G. A., *Lectures on the Calculus of Variations*, The University of Chicago Press, Chicago, 1946.
BOLTYANSKII, V. G., *Mathematical Methods of Optimal Control*, Holt, Rinehart and Winston, New York, 1971.
BOLZA, O., *Lectures on the Calculus of Variations*, Dover Publications, New York, 1961.
BRYSON, A. E., Jr., and HO, Y. C., *Applied Optimal Control*, Blaisdell Publishing Company, Lexington, Massachusetts, 1969.
BURGHES, D. N., and GRAHAM, A., *Introduction to Control Theory Including Optimal Control*, John Wiley and Sons, New York, 1980.
CANON, M. D., CULLUM, D. C., Jr., and POLAK, E., *Theory of Optimal Control and Mathematical Programming*, McGraw-Hill Book Company, New York, 1970.
CHANG, S. S. L., *Synthesis of Optimum Control Systems*, McGraw-Hill Book Company, New York, 1961.

CICALA, P., *An Engineering Approach to the Calculus of Variations*, Levrotto e Bella, Turin, 1957.
CITRON, S. J., *Elements of Optimal Control*, Holt, Rinehart and Winston, New York, 1969.
CONVERSE, A. O., *Optimization*, Holt Rinehart and Winston, New York, 1970.
DENN, M. M., *Optimization by Variational Methods*, McGraw-Hill Book Company, New York, 1969.
DENNIS, J., *Mathematical Programming and Electrical Networks*, John Wiley and Sons, New York, 1959.
DREYFUS, S. E., *Dynamic Programming and the Calculus of Variations*, Academic Press, New York, 1965.
DYER, P., and MCREYNOLDS, S. R., *The Computation and Theory of Optimal Control*, Academic Press, New York, 1970.
ELSGOLC, L. E., *Calculus of Variations*, Addison-Wesley Publishing Company, Reading, Massachusetts, 1962.
EVELEIGH, V. W., *Adaptive Control and Optimization Techniques*, McGraw-Hill Book Company, New York, 1967.
EWING, G. M., *Calculus of Variations with Applications*, W. W. Norton and Company, New York, 1969.
FAN, L. T., *The Continuous Maximum Principle*, John Wiley and Sons, New York, 1966.
FAN, L. T., and WANG, C. S., *The Discrete Maximum Principle*, John Wiley and Sons, New York, 1964.
FELDBAUM, A. A., *Optimal Control Systems*, Academic Press, New York, 1965.
FLÜGGE-LOTZ, I., *Discontinuous Automatic Control*, Princeton University Press, Princeton, New Jersey, 1953.
FLÜGGE-LOTZ, I., *Discontinuous and Optimal Control*, McGraw-Hill Book Company, New York, 1968.
FORRAY, M. J., *Variational Calculus in Science and Engineering*, McGraw-Hill Book Company, New York, 1968.
GELFAND, I. M., and FOMIN, S. V., *Calculus of Variations*, Prentice-Hall, Englewood Cliffs, New Jersey, 1963.
GIBSON, J. E., *Nonlinear Automatic Control*, McGraw-Hill Book Company, New York, 1963.
HERMANN, R., *Differential Geometry and the Calculus of Variations*, Academic Press, New York, 1968.
HERMES, H., and LASALLE, J. P., *Functional Analysis and Time Optimal Control*, Academic Press, New York, 1969.
HESTENES, M. R., *Calculus of Variations and Optimal Control Theory*, John Wiley and Sons, New York, 1966.
INTRILIGATOR, M. D., *Mathematical Optimization and Economic Theory*, Prentice-Hall, Englewood Cliffs, New Jersey, 1971.
JACOBSON, D. H., *Extensions of Linear-Quadratic Control, Optimization and Matrix Theory*, Academic Press, New York, 1977.
JACOBSON, D. H., and MAYNE, D. Q., *Differential Dynamic Programming*, American Elsevier Publishing Company, New York, 1970.
KAMIEN, M. I., and SCHWARTZ, N. L., *Dynamic Optimization*, Elsevier Publishing Company, New York, 1981.
KIPINIAK, W., *Dynamic Optimization and Control*, John Wiley and Sons, New York, 1961.
KIRK, D. E., *Optimal Control Theory: An Introduction*, Prentice-Hall, Englewood Cliffs, New Jersey, 1970.
LAPIDUS, L., and LUUS, R., *Optimal Control of Engineering Processes*, Blaisdell Publishing Company, Waltham, Massachusetts, 1967.

Bibliography

LAWDEN, D. F., *Optimal Trajectories for Space Navigation*, Butterworth and Co., London, 1963.
LEE, E. B., and MARKUS, L., *Foundations of Optimal Control Theory*, John Wiley and Sons, New York, 1967.
LEITMANN, G. (editor), *Optimization Techniques*, Academic Press, New York, 1962.
LEITMANN, G. (editor), *Topics in Optimization*, Academic Press, New York, 1966.
LEITMANN, G., *An Introduction to Optimal Control*, McGraw-Hill Book Company, New York, 1966.
LUENBERGER, D. G., *Optimization by Vector Space Methods*, John Wiley and Sons, 1969.
MCCAUSLAND, I., *Introduction to Optimal Control*, John Wiley and Sons, New York, 1969.
MAREC, J. P., *Optimal Space Trajectories*, Elsevier Publishing Company, New York, 1979.
MERRIAM, C. W., III, *Optimization Theory and the Design of Feedback Control Systems*, McGraw-Hill Book Company, New York, 1964.
MIELE, A. (editor), *Theory of Optimum Aerodynamic Shapes*, Academic Press, New York, 1965.
OGUZTÖRELI, M. N., *Time-Lag Control Systems*, Academic Press, New York, 1966.
OLDENBURGER, R., *Optimal Control*, Holt, Rinehart and Winston, New York, 1966.
PALLU DE LA BARRIÈRE, R., *Optimal Control Theory*, W. B. Saunders Company, Philadelphia, 1967.
PETERSON, E. L., *Statistical Analysis and Optimization of Systems*, John Wiley and Sons, New York, 1961.
PETROV, I. P., *Variational Methods in Optimum Control Theory*, Academic Press, New York, 1968.
PIERRE, D. A., *Optimization Theory and its Applications*, John Wiley and Sons, New York, 1969.
POLAK, E., *Computational Methods in Optimization*, Academic Press, New York, 1971.
PONTRYAGIN, L. S., BOLTYANSKII, V. G., GAMKRELIDZE, R. V., and MISHCHENKO, E. F., *The Mathematical Theory of Optimal Processes*, Interscience Publishers, New York, 1962.
RUND, H., *The Hamilton–Jacobi Theory in the Calculus of Variations*, Van Nostrand Company, New York, 1966.
SAGAN, H., *Introduction to the Calculus of Variations*, McGraw-Hill Book Company, New York, 1969.
SAGE, A. P., and WHITE, C. M., *Optimum Systems Control*, Prentice-Hall, Englewood Cliffs, New Jersey, 1977.
SALUKVADZE, M. E., *Vector-Valued Optimization Problems in Control Theory*, Academic Press, New York, 1979.
SCHECHTER, R. S., *The Variational Method in Engineering*, McGraw-Hill Book Company, New York, 1967.
TABAK, D. and KUO, B. C., *Optimal Control by Mathematical Programming*, Prentice-Hall, Englewood Cliffs, New Jersey, 1971.
TOU, J. T., *Modern Control Theory*, McGraw-Hill Book Company, New York, 1964.
TRUXAL, J., *Automatic Feedback Control Systems Synthesis*, McGraw-Hill Book Company, New York, 1955.
TSIEN, H. S., *Engineering Cybernetics*, McGraw-Hill Book Company, New York, 1954.
VARAIYA, P. P., *Notes on Optimization*, Van Nostrand Reinhold Co., New York, 1972.
VINCENT, T. L., and GRANTHAM, W., *Optimality in Parametric Systems*, John Wiley and Sons, New York, 1981.
WILDE, D. J. and BEIGHTLER, C. S., *Foundations of Optimization*, Prentice-Hall, New York, 1967.
YOUNG, L. C., *Calculus of Variations and Optimal Control Theoy*, Saunders, Philadelphia, 1969.
ZAHRADNIK, R. L., *Theory and Techniques of Optimization*, Barnes and Noble, New York, 1971.

Index

Accessory minimum problem, 55
Adjoint, 112
   equations, 112, 119, 174
Autonomous system, 82

Bang–bang control, 213
   number of switches, 215
   for time-optimality, 215
Basis, xii
Blaquière, Austin, v
Bolza, Oskar, v

Calculus of variations
   inverse problem, 39
   simplest problem of, 7, 164
Conjugate point, 58
Conjugate value, 58, 61
Constraints, 28, 31, 71, 77
   differential, 72
   end point inequality, 196
   finite, 72
   isoperimetric, 30, 181
   state-dependent, 168
Control, 77
   admissible, 81
   admissible at x^0, 168, 182
   constraints, 78, 168
   extremal at x^0, 121
   feasible at x^0, 81, 168, 182
   feedback, 266
   joining of, 83
   optimal at x^0, 82, 265
   optimal at x^0 with respect to X, 242
   optimal on θ^0, 141

Control (*continued*)
   optimal on θ^0 with respect to X, 244
   Pareto-optimal at x^0, 293
   singular, 211, 225, 233
   time-optimal, 125, 130, 174, 213, 217, 247
   variables, 77, 79
Corless, Martin, vi
Corner, xiii
   conditions, 67, 68
   lemma, 56
Cost, 81
   additivity of, 85
   optimal, 85
   terminal, 145
   variable, 88
   vector-valued, 285
Criterion, 81

Dido's problem, v, 30, 188
Distance
   of order 0, 11
   of order 1, 11
DuBois-Reymond lemma, 14
Dynamic programming, 121

Efficient decision, 292
End points
   fixed, 7
   variable, 72
Erdmann–Weierstrass corner conditions, 67
Euler, Leonhard, 15
   necessary condition of, 15

Euler–Lagrange equation, 17
   finite equation, 33
   identity, 33, 35
   integration of, 25
Ewing, George M., v
Extremals, 19
   with corners, 68, 69

Feedback control, 265
   discontinuous, 276, 283
   existence of, 275
   optimal, 267
   synthesis of, 268
Field theorem, 242, 246
Function, xiii
   admissible, 7
   of class C^k, xiii
   convex, xiv
   global (absolute) minimum of, 8
   local (relative) minimum of, 9
   piecewise continuous, xiii
   piecewise smooth, xiii
   restriction of, xiv
   smooth, xiii
   vector-valued, xiv, 21
Functional, 8, 82

Hamilton's principle, 43
Hamilton–Jacobi–Bellman equation, 269

Indifference curve, 286

Jacobi, C. G. J., 55
   necessary condition of, 58
Jacobi's equation, 57
   integration of, 60
   solutions of, 62

Lagrange, J. L., 17
Legendre, A. M., 51
   necessary condition of, 51
Leibniz, G. W., 16
Limiting surface, 90
   fundamental property of, 92
   regular interior point of, 99
   tangent plane of, 99
   transfer of tangent plane of, 110
Linear-quadratic problem, 269
Luenberger, David G., vi

Maximum principle, 118

Minimum
   local, 12
   necessary condition for, 13, 17, 18, 21, 47, 51, 58
   strong local, 12
   weak local, 12

Neighborhood, 9
   of order 0, 11
   of order 1, 11
Nonautonomous system, 155
Notation, xi

Optimal control, 82
   bang–bang, 128, 162, 213, 215
   closed-loop, 267
   feedback, 267
   necessary conditions for, 105, 150, 160, 187
   simplest problem of, 82
   sufficient conditions for, 242, 246, 250
Optimal isocost surface, 90
Optimality, principle of, 86
Ordering, 288

Parameter optimization, 190
   necessary conditions for, 194
Pareto, V., 291
Pareto-optimal control, 292
   necessary conditions for, 293, 294
   sufficient conditions for, 295
Pareto-optimal decision, 292
Performance index, 82
Preference, 288

Regular interior point, 99
   necessary conditions at, 105
Regulator, time-optimal, 217, 222

Schmitendorf, William E., vi
Set, closure of, xii
Spaces, xii
Stadler, Wolfram, vi
State, 77
   augmented, 88
   measurement error, 277
   variables, 77, 79
State equations, 77, 80, 119
   autonomous, 81
   linear, time-invariant, 211

Index

Stationarity, 22
    necessary and sufficient condition for, 23
    principles, 43
Sufficient conditions, 241, 242
    direct, 249
Switching curve, 219, 265
Switching function, 212, 226
Symbols, xi

Target set, 81
    smooth manifold, 115
Trajectory, 89
    fundamental property of, 92
    optimal, 89
    perturbed, 266, 277, 283
    regular optimal, 111
    terminating, 89

Transversality condition
    initial, 140
    terminal, 115

Variation
    first, 22
    total, 22
Variational equations, 107, 173
Vector, xii
Vector-valued cost, 285, 292
Vincent, Thomas L., vi

Weierstrass, Karl, 47
Weierstrass
    excess function, 47
    necessary condition of, 47